D0416305

ELECTRICITY Made Simple

The Made Simple series
has been created
primarily for self-education
but can equally well
be used as
an aid to group study.
However complex the subject,
the reader is taken,
step by step,
clearly and methodically
through the course. Each volume
has been prepared by
experts,
using throughout the
Made Simple technique of teaching.
Consequently the gaining
of knowledge now becomes
an experience to be enjoyed.

In the same series

Accounting
Acting and Stagecraft
Additional Mathematics
Advertising
Anthropology
Applied Economics
Applied Mathematics
Applied Mechanics
Art Appreciation
Art of Speaking
Art of Writing
Biology
Book-keeping
British Constitution
Business and Administrative Organisation
Calculus
Chemistry
Childcare
Commerce
Company Administration
Company Law
Computer Programming
Cookery
Cost and Management Accounting
Data Processing
Dressmaking
Economic History
Economic and Social Geography
Economics
Electricity
Electronic Computers
Electronics
English
English Literature
Export
Financial Management
French
Geology
German
Human Anatomy
Italian
Journalism
Latin
Law
Management
Marketing
Mathematics
Modern Electronics
Modern European History
New Mathematics
Office Practice
Organic Chemistry
Philosophy
Photography
Physical Geography
Physics
Pottery
Psychology
Rapid Reading
Retailing
Russian
Salesmanship
Secretarial Practice
Social Services
Soft Furnishing
Spanish
Statistics
Transport and Distribution
Typing
Woodwork

ELECTRICITY Made Simple

Leslie Basford, B.Sc.

Made Simple Books
W. H. ALLEN London
A Howard & Wyndham Company

© W. H. Allen & Co. Ltd., 1968

Made and printed in Great Britain
by Richard Clay (The Chaucer Press), Ltd., Bungay, Suffolk
for the publishers W. H. Allen & Co. Ltd.,
44 Hill Street, London W1X 8LB

First published October 1968
Reprinted, May 1971
Reprinted, October 1972
Reprinted, August 1974
Reprinted (with Revision Questions), April 1976
Reprinted, February 1978
Reprinted, January 1979

This book is sold subject to the
condition that it shall not, by
way of trade or otherwise, be lent,
re-sold, hired out, or otherwise
circulated without the publisher's
prior consent in any form of binding
or cover other than that in which it is
published and without a similar condition
including this condition being imposed
on the subsequent purchaser

ISBN 0 491 01741 3 Paperbound

Preface

Electricity Made Simple gives the reader a knowledge of the fundamental principles underlying every aspect of the subject. It is particularly valuable as a background course for the student who intends to continue his studies to a professional level, either into electrical engineering or electronics. For the enterprising lay student who wants to make use of this knowledge at home—in the understanding of domestic appliances, for instance—the book will give all the answers and more. It will also enable the intelligent parent to keep abreast of his children's school work or to revise his own forgotten knowledge.

The book goes well beyond the requirements of G.C.E. O-level and much of it will apply to post-O.N.C. students too. Each aspect of electricity is considered from its basic ideas and a large number of worked examples help the reader to assimilate the topic under discussion. Where the text deals with more technical material, that, for instance, concerning alternating-current theory in the second half of the book, the student will be helped by the glossary of terms at the end. The rationalized M.K.S. system of units is employed, but parallel examples of the C.G.S. system are included for the benefit of readers who used this latter system at school.

The author is extremely grateful to Professor M. G. Say for reading the manuscript, and wishes to record his sincere thanks for his invariably courteous and helpful suggestions.

Table of Contents

CHAPTER ONE

CURRENT

In a civilization where homes without television, electric light, radio, or refrigerator are the exception rather than the rule, it is none the less true that the average consumer has at best a very sketchy idea of what he is paying for each time his electricity bill falls due.

Questioned about the nature of electricity, most people will acknowledge that it is a useful commodity; some will concede that it is a 'current'; a few will volunteer that an electric current is a flow of 'electrons'; one or two will know that electrons are parts of atoms. So an understanding of electricity must begin with the atom.

All matter is made up of atoms—of which ninety-two varieties occur naturally, and about a further dozen have been made artificially in the laboratory. Any atom is an unimaginably small particle, and is itself made up from even smaller particles, including electrons. According to a rather oversimplified picture, we may imagine that an atom is like the solar system, with the main mass concentrated in the central nucleus (corresponding to the Sun), around which the electrons (corresponding to the planets) move in orbits.

The important feature of this picture is that the central nucleus is charged with electricity: it carries what is conventionally called a *positive* charge. Yet the atom as a whole is not charged with electricity: it is electrically neutral because the positive charge on its nucleus is balanced by the opposite charges on the orbiting electrons. So each electron carries what is conventionally called a *negative* charge. (The convention of naming charges was devised long before the electron was discovered. It is a pity, as we shall see later, that it did not happen the other way round.)

ELECTRONS

The outermost electrons in the solar-system picture of the atom are held in orbit much less securely than the inner electrons. Thus the outermost electron—in the N shell—of a copper atom is easily dislodged from its very elliptical orbit (Fig. 1), e.g. by collision. Electrons that become dislodged can exist by themselves quite independently of their parent atoms, and are then known as *free electrons*.

All solids contain some free electrons. Substances which contain a lot are good electrical conductors (e.g. copper, silver, gold), while substances which contain very few free electrons are bad electrical conductors and are known as insulators (e.g. rubber, glass, polythene).

Apart from the fact that it may become hot, no apparent change occurs in a metal when it is carrying a current. The atoms are fixed in a regular pattern, or lattice, and the free electrons drift about in the empty space between them (Fig. 2).

Normally the free electrons drift in a random manner, so that, on balance, there are just as many moving in one direction as in any other. If, however, the free electrons can be organized to drift in a definite direction the result will be a movement of electricity through the conductor—in other words, a *current* of electricity is set up.

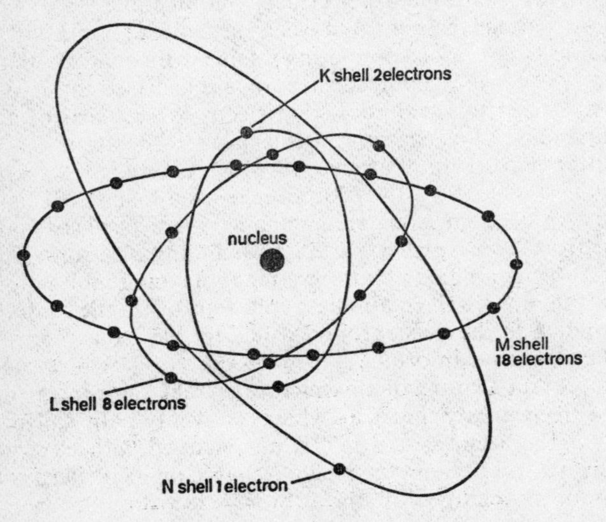

Fig. 1. Simplified model of a copper atom; only one orbit has been drawn for each 'shell' of electrons

(Although the atoms in a solid are fixed, the atoms making up a liquid or a gas are free to move around. As we shall discover when we come to electrochemistry, the atoms in certain liquids and gases not only move they also carry electrical charges. Such charged atoms are called *ions*, and they owe their charges either to gaining one or more electrons or to losing one or more electrons. Hence, to avoid confusion later on, we ought to bear in mind that in conducting fluids, as opposed to conducting solids, current is carried by ions.)

So a current, or to be precise, a current in a metallic conductor, is a flow of electrons. Furthermore, the rate of flow, i.e. the number of electrons passing any point in one second, gives the size of the current.

As we saw earlier, an electron is a very small particle and carries a very small charge of electricity; a colossal number of electrons has to flow before we have a measurable current.

A flow of some $6{\cdot}28 \times 10^{18}$ (six million two hundred and eighty

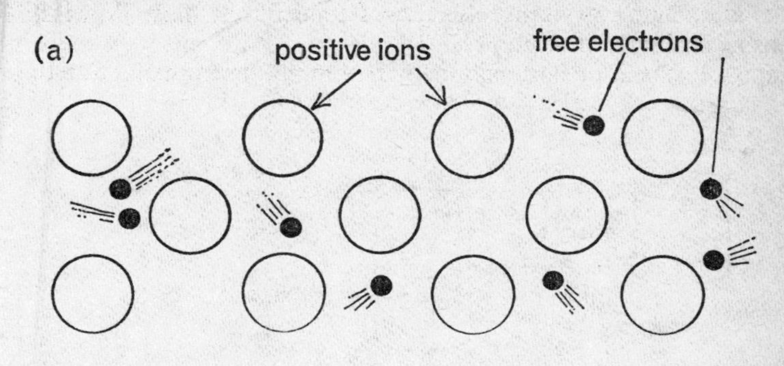

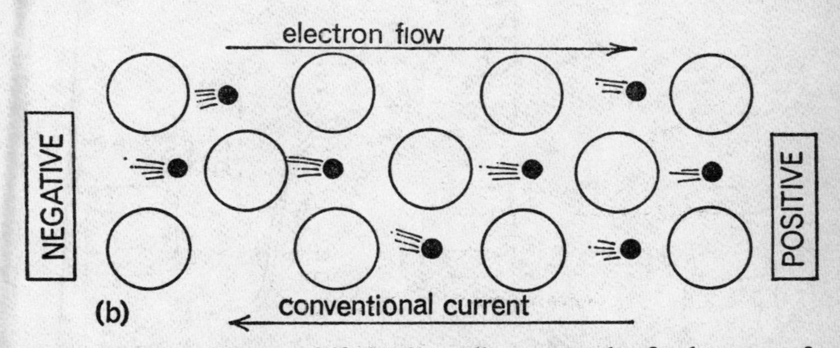

Fig. 2. (*a*) Free electrons drift haphazardly among the fixed atoms of a conductor. (*b*) An organized drift in one direction only constitutes a current of electricity

thousand *billion*) electrons per second is a current of one ampere or 'amp'—the practical unit of current.

CURRENT AND VOLTAGE

Anyone who has ever changed a fuse will have encountered the term 'amp', but what, one may justifiably ask, is the difference between it and the 'volt'? What exactly is meant by a '12-volt' car battery or a '240-volt' supply? And what is this thing that one is warned to avoid by 'danger, high voltage' notices?

To get some idea of what voltage means and how it is related to current, let us imagine that the flow of electrons through a conducting wire is like the flow of water through a pipe. Fig. 3 shows water flowing naturally down a sloping pipe. The rate of flow varies with the difference in height between the ends of the pipe: if there is no height difference there is no flow.

Water, like everything else, has a tendency to flow downhill. The reason is simply that its *potential energy* (i.e. the energy stored in the water by virtue of its position) is *less* at the bottom than at the top.

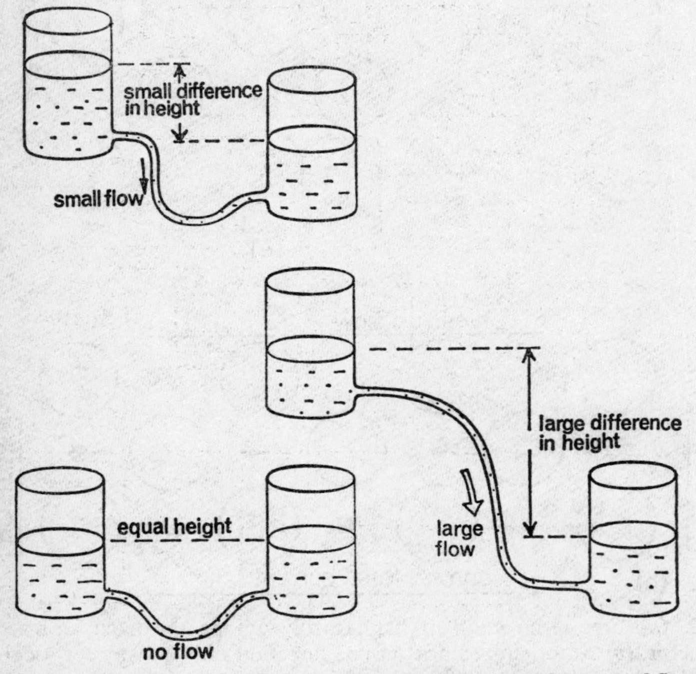

Fig. 3. Water flows naturally down a sloping pipe; the rate of flow depends on the difference in height between the ends of the pipe

Electrons share this tendency to move if they possibly can to the position where their potential energy is least.

The difference in height between the ends of the water pipe is a measure of the difference in potential energy, and, as Fig. 3 indicates, it is this potential-energy difference that controls the rate of flow. Similarly, it is the difference of potential energy ('potential difference' for short) between the ends of a conductor that governs the flow of electrons (current) through it. If there is no potential difference there is no current.

There is nothing alarming about the expression 'potential difference': it is merely the proper name for what is commonly called 'voltage'. (It

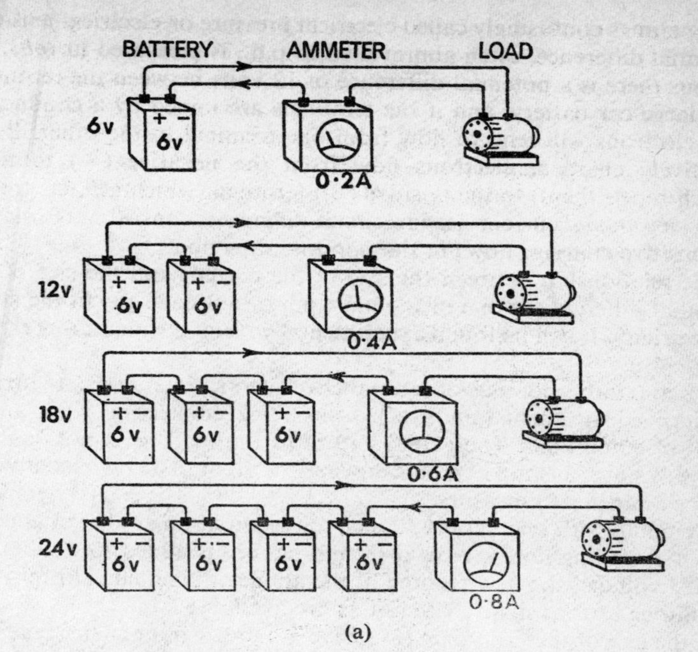

(a)

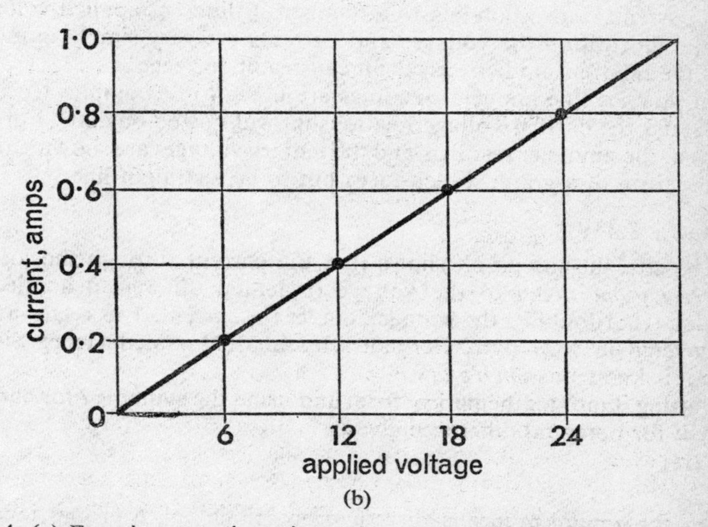

(b)

Fig. 4. (*a*) Experiment to investigate the relationship between current and potential difference. (*b*) Result plotted as a graph is a straight line

is sometimes confusingly called electrical pressure or electrical tension.) Potential difference, often abbreviated to p.d., is measured in *volts*.

Thus there is a potential difference of 12 volts between the terminals of a large car battery, and if the terminals are joined by a conducting wire electrons will tend to flow from one terminal to the other. Being negatively charged, electrons flow from the negative (−) terminal (which repels them) to the positive (+) terminal (which attracts them). The *conventional* current used in circuit diagrams, consisting of imaginary positive charges, flows in the opposite direction.

The relationship between the size of the current and the size of the potential difference is one of the most important concepts in the study of electricity. It can be found experimentally using the apparatus shown in Fig. 4.

The apparatus consists of a number of identical batteries, a current-measuring instrument (ammeter), some thick connecting wire, and a 'load' of some kind. The actual nature of the load need not concern us at this stage, as long as it cuts down the current to a convenient value for the ammeter to measure.

Beginning with one battery, the circuit is wired up as shown and the ammeter reading noted. Now a second battery is added to double the applied voltage; the new reading of the ammeter is noted. This process is continued for as many batteries as are available.

At the end of the experiment we have a number of ammeter readings (i.e. currents) corresponding to a number of different applied voltages. In the illustration the voltages and currents are very simple numbers, and the relationship between them can readily be seen.

In practice, the ammeter readings are likely to be complicated decimals and the current/voltage relationship will not be obvious. For this reason, the ammeter readings and the battery voltages are shown plotted in the form of a graph, which turns out to be a straight line.

OHM'S LAW

The straight-line graph shows that the current through the load is *directly proportional* to the voltage (potential difference) applied to the load, i.e. doubling the voltage doubles the current. This relationship, discovered in 1827 by a German schoolmaster named Georg Simon Ohm, is known as *Ohm's law*.

Putting it into mathematical form and using the symbols I for current and V for potential difference gives

$$I \propto V$$

where the symbol $\propto$ means 'directly proportional to'. A proportionality such as this can be written more conveniently in the form of an equation:

$$I \times \text{constant} = V$$

The constant term in the above equation is a measure of the opposition that the 'load' offers to the current passing through it, and is known as *resistance.*

Using the symbol R for resistance, we can write Ohm's law in the form

$$V = IR$$

or

$$I = \frac{V}{R}$$

or

$$R = \frac{V}{I}$$

When, as is normally the case, current is measured in amperes and the potential difference is measured in volts the resistance is measured in units called *ohms*, whose symbol is the Greek letter omega: Ω. (The inverse or opposite of resistance is *conductance*. Using the symbol G for conductance, Ohm's law can be written $G = \frac{1}{R} = \frac{I}{V}$. It is measured in 'reciprocal ohms' or siemens.)

In the experiment shown in Fig. 4, the single 6-V battery drives a current of 0·2 A through the load. We now know from Ohm's law that

$$R \text{ (ohms)} = V \text{ (volts)} \div I \text{ (amps)}$$

so we can calculate the resistance of the load. It is

$$6 \text{ (volts)} \div 0{\cdot}2 \text{ (amps)} = 30\ \Omega$$

We have assumed that only the load offers resistance to the current. In fact, the batteries, ammeter, and even the connecting wires have some resistance. But as this is comparatively small (certainly less than 1 Ω), we may justifiably ignore it.

Ohm's law is not a universal law but an experimental fact that applies to certain types of conductor. It applies to pure metals and metallic alloys (for a given steady temperature), and these are known as linear conductors, since they—like the load in Fig. 4—give a straight-line graph for the relationship between voltage and current. Materials which give a curved or angular graph are known as non-linear conductors; they do not obey Ohm's law. Examples of a few non-linear conductors are listed in Fig. 5.

Fig. 5 (*a*) is an example of an ideal conductor which obeys Ohm's law at all times, even when the current is reversed. Fig. 5 (*b*) shows how a semiconductor such as germanium offers a low resistance to currents flowing in one direction, but offers a much higher resistance when the current is reversed. Fig. 5 (*c*) is the characteristic curve of a thermionic valve: Ohm's law applies to the valve only if operated in the straight portion of its characteristic. Fig. 5 (*d*) is the current/voltage

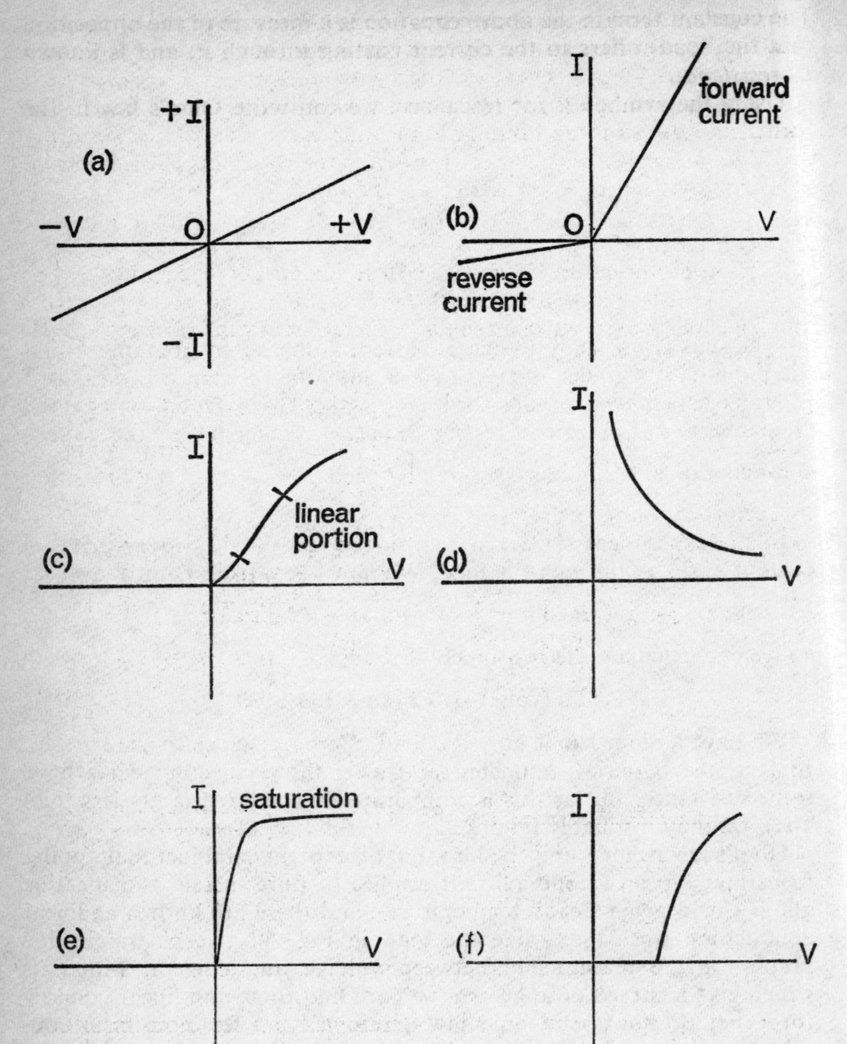

Fig. 5. Not all conductors obey Ohm's law. (*a*) Ideal conductor. (*b*) Semi-conductor. (*c*) Thermionic valve. (*d*) Carbon-arc lamp. (*e*) Gas-filled tube. (*f*) Electrolyte

relationship for a carbon-arc lamp: surprisingly, the current drops as the voltage is increased; this is described as a 'negative resistance' characteristic. Fig. 5 (*e*) shows how a gas such as neon in an advertising sign obeys Ohm's law for small potential difference, while large p.d.s produce 'saturation' and the current is unaffected by increasing voltages.

In some of these non-linear conductors the current is carried by *ions*, i.e. charged atoms, rather than free electrons.

SIMPLE CIRCUITS

Before going on to discuss electrical circuits, it is helpful to turn again to the water-pipe analogy. The first part of Fig. 6 is very similar to Fig. 3 on p. 4, except that a turbine or water-wheel has been

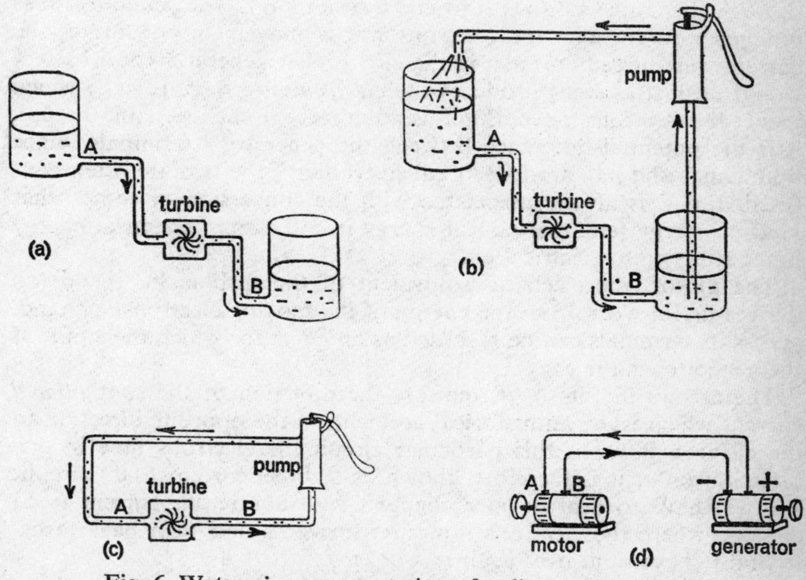

Fig. 6. Water-pipe representation of a direct-current circuit

inserted into the system. As the water flows from the upper tank to the lower it drives the turbine, and useful work could be obtained from it. The potential energy of the water is turned into kinetic energy (energy of motion). The turbine is simply a device for deriving useful work from the kinetic energy of the water.

Plainly the simple system shown in Fig. 6 (*a*) would not prove very practicable, since it comes to a halt as soon as the water stored in the upper tank is exhausted. To avoid this, it is necessary to replace the water just as quickly as it runs out. But water does not flow uphill of its own accord; we therefore need some kind of pump, as shown in Fig. 6 (*b*). There is now no need for the two tanks, and they can be removed (Fig. 6 (*c*)).

The water in Fig. 6 (*c*) is flowing around a closed path from pump to point A, through the turbine to point B, and back to the pump: the pipes form a complete *circuit*.

Work done at the pump handle reappears as useful work obtainable from the turbine. Some of the work done, however, is wasted in driving water through the pipes and through the various valves, etc., of the pump and turbine; the waste may be very considerable if the pipes are narrow or very long.

The electrical equivalent of this circuit is shown in Fig. 6 (*d*), where a generator or dynamo takes the place of the pump in forcing electrons through the connecting wires to an electric motor. The generator does not generate current: the electrons whose movement constitutes the current were already present in the circuit. The generator (or a battery would do just as well) produces the *electromotive force* (e.m.f.) which makes the electrons move. E.m.f. is not a force in the mechanical sense, it is the potential difference between the generator's terminals. Since both e.m.f. and p.d. are loosely called 'voltage', the two are often confused. E.m.f. is always associated with the conversion of some other kind of energy (e.g. mechanical energy in a dynamo, chemical energy in a battery) into electrical energy.

The motor, the electrical equivalent of the turbine in Fig. 6 (*c*), derives useful work from the energy of the moving electrons. The p.d. across its terminals can be regarded as an *effect* for which the e.m.f. of the generator is the *cause*.

The arrows in Fig. 6 (*d*) indicate the direction of the *conventional* current which is, by an historical accident, in the opposite direction to the electron flow. In this particular circuit the electrons flow in one direction only; it is therefore known as a direct-current (d.c.) circuit. We will have to deal in later chapters with alternating-current (a.c.) circuits, where the electrons flow first forwards and then backwards, to and fro over and over again.

SUMMARY

Electrons are minute negatively charged particles which move in orbits around the central nucleus of an atom. Free electrons exist independently of the atoms from which they have been dislodged. An electric current is an organized drift of free electrons: the size of the current is the *rate* of flow of electrons.

Metals contain relatively large numbers of free electrons, and are therefore good conductors of electricity. Insulating materials contain very few free electrons.

Potential difference (p.d.) or 'voltage' is a measure of the difference in potential energy between two points. Electrons naturally tend to flow towards the point where their potential energy is least, i.e. the least negative part of a conductor, which is, of course, the same as the most positive part.

Current is measured in amperes (amps). Potential difference is measured in volts.

The relationship between current I and potential difference V is given by Ohm's law: the size of the current through a conductor is directly proportional to the potential difference between its ends.

$$I = \frac{V}{R}$$

where R is the *resistance* of the conductor. Resistance is measured in ohms. Not all conductors obey Ohm's law.

The potential difference between the terminals of a battery or generator when it is on 'open circuit', i.e. not delivering any current, is known as an electromotive force (e.m.f.).

Conventional current, as used in circuit diagrams, is in the opposite direction to the actual flow of electrons.

CHAPTER TWO

RESISTANCE

Any circuit that carries a current has to comply with two conditions: it must contain a potential difference, and it must provide an unbroken path for the electrons to traverse. Both conditions are fulfilled by the simple circuit shown in Fig. 7.

The circuit diagram (Fig. 7 (*b*)) is a shorthand method of depicting

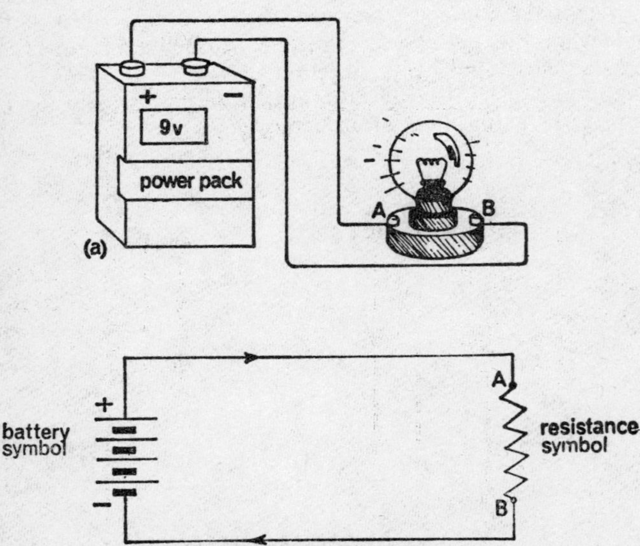

Fig. 7. (*a*) Battery and bulb circuit. (*b*) Equivalent circuit diagram

all we need to know about the battery-and-bulb arrangement of Fig. 7 (*a*).

Electrons flow from the negative terminal of the battery (in the opposite direction to the conventional current indicated by arrows on the circuit diagram), and have no difficulty in passing through the relatively thick copper wire to terminal B of the light bulb. When they reach the filament, which is a very thin tungsten wire, the flow is impeded; the filament offers *resistance* to the current. Resistance is indicated by a zig-zag symbol in the circuit diagram.

The electrons lose some of their energy in passing through the filament. What happens in fact is that the energy of the electrons is transferred to the tungsten atoms, which vibrate more and more violently until the filament is white hot.

In general, any conductor that offers resistance to current converts electrical energy into heat energy. This can be exploited, as described in Chapter Three, or it can be a nuisance, as in an electric motor, where expensive ventilating facilities have to be provided to prevent overheating.

INTERNAL RESISTANCE

According to the label on the battery in Fig. 7, it can deliver 9 V. This figure refers to the e.m.f. or 'open-circuit voltage', and is the

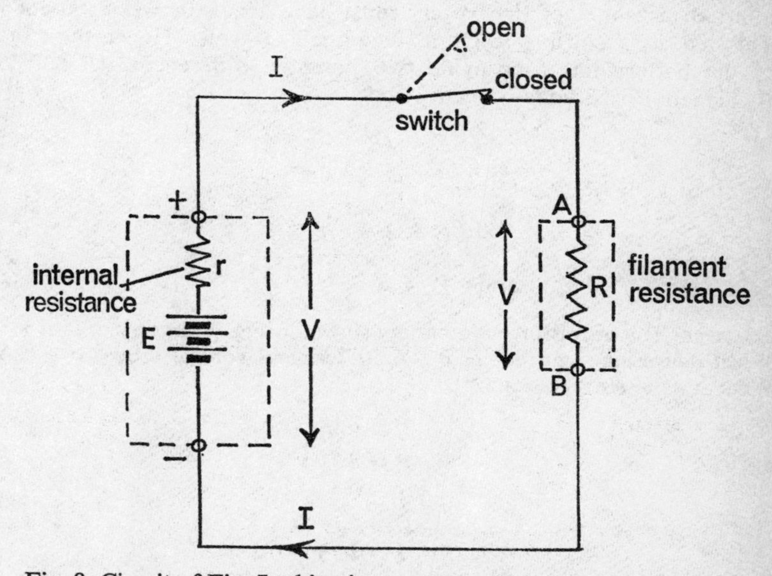

Fig. 8. Circuit of Fig. 7 taking into account the internal resistance of the battery

voltage between the battery terminals when no current is flowing. When current is drawn from the battery the voltage between the terminals is appreciably less than 9 V. The reason for this drop in voltage is that the battery itself offers resistance to the current in the circuit. If the battery were replaced by a dynamo, that, too, would offer resistance.

Fig. 8 shows the battery-and-bulb circuit again, but this time account has been taken of the battery's own opposition to current—called internal resistance. There are now two resistances in the circuit: that of the lamp filament is R ohms and the internal resistance of the battery is r ohms. With the switch open so that no current can flow the voltage between the terminals of the battery is its e.m.f. E volts.

With the switch closed, a current of I amps is drawn from the battery and the voltage between its terminals drops to V volts.

Since each terminal of the battery is now connected directly to the lamp, the p.d. between the lamp terminals is also V volts. If the resistance of the lamp is R ohms and the current passing through it is I amps we know from Ohm's law that the p.d. between its terminals is IR volts. Hence

$$V = IR$$

The same current of I amps passes also through the battery, so the internal resistance of the battery must have a p.d. between its 'ends'. This p.d. is, according to Ohm's law again, Ir volts. Hence the e.m.f. of the battery has to provide two potential differences, IR and Ir. If this e.m.f. is E volts we can write

$$E = IR + Ir$$

or, since $V = IR$,

$$E = V + Ir$$

$$\text{and} \quad V = E - Ir$$

EXAMPLE: The transistor-radio battery shown in Fig. 7 has an e.m.f. of 9 V. When delivering a current of 0·3 A its terminal voltage drops to 8·76 V. What is its internal resistance?

$$\begin{aligned} E &= 9\text{ V} \\ V &= 8{\cdot}76\text{ V} \\ I &= 0{\cdot}3\text{ A} \\ E &= V + Ir \\ 9 &= 8{\cdot}76 + 0{\cdot}3 \times r \\ 0{\cdot}3 \times r &= 0{\cdot}24 \\ r &= \frac{0{\cdot}24}{0{\cdot}3} \\ &= 0{\cdot}8\ \Omega \quad \textit{Answer} \end{aligned}$$

The resistance of the tungsten filament in the battery-and-bulb circuit that we have been considering is not steady. Its value is 15–20 times as high when white hot as when the bulb is cold. Consequently, as the voltage applied to the lamp is increased from zero to its maximum value, the resistance of the filament increases almost as quickly, and the current ($I = V/R$) remains practically constant. This does not mean that tungsten wire violates Ohm's law, it is simply that Ohm's law applies only under *steady* conditions of temperature and pressure.

EFFECT OF TEMPERATURE

Metallic conductors all show this increase of resistance with temperature, though by differing amounts. As the temperature goes up, their resistance increases; as the temperature goes down, their resistance decreases. The increase in resistance is (approximately) directly proportional to the increase in temperature.

Thus, if the resistance of a conductor is R_0 ohms at 0° C and its resistance at a temperature of t° C is R_t ohms, the increase in resistance $(R_t - R_0)$ is directly proportional to t. Mathematically,

$$(R_t - R_0) \propto t$$

or, in the more useful form of an equation,

$$R_t - R_0 = R_0 \alpha t$$

where α (the Greek letter alpha) is a constant called the 'temperature coefficient of resistance'. Rearranging the equation gives

$$R_t = R_0(1 + \alpha t)$$

This enables us to calculate the resistance of a conductor at any temperature if we know its resistance at 0° C.

In practice, the resistance would be measured at the temperature of the laboratory (about 20° C) rather than at 0° C. The same equation is still valid with reasonable accuracy if R_0 is the measured resistance at some temperature other than 0° C; the symbol t then stands for the *increase* in temperature (i.e. final temperature less the original temperature).

The temperature coefficient of resistance depends on the material from which the conductor is made. Typical values for the more common electrical conductors are given in Table 1.

Table 1. TEMPERATURE COEFFICIENTS OF RESISTANCE (PER DEG C)

Material	α
Copper	0·00428
Aluminium	0·00435
Iron	0·00625
Manganin (Cu + Mn + Ni)	0·00001
Carbon	−0·0005

EXAMPLE: An iron wire has a resistance of 50 Ω at 20° C. What is its resistance at 100° C?

The increase in temperature is 100° − 20° = 80° C

$$
\begin{aligned}
t &= 80^\circ \text{ C} \\
R_0 &= 50\ \Omega \\
\alpha &= 0{\cdot}00625 \text{ per deg C (from Table 1)} \\
R_t &= R_0(1 + \alpha t) \\
&= 50\ (1 + 0{\cdot}00625 \times 80) \\
&= 50\ (1{\cdot}5) \\
&= 75\ \Omega \quad \textit{Answer}
\end{aligned}
$$

Resistance can be measured with great accuracy (using the Wheatstone bridge described on p. 32), and the increase in resistance of a conductor as its temperature increases may therefore be used for comparing temperatures. This is the principle of the *resistance thermometer*, which uses a coil of fine platinum wire. The resistance of the coil is found at 0° C and at the temperature to be measured: if these resistances are R_0 and R_t respectively the temperature to be measured (t) is given by $R_t = R_0(1 + \alpha t + \beta t^2)$, a more exact form of the equation derived earlier.

In Table 1 the temperature coefficient of carbon is given as *minus* 0·0005 per deg C. Carbon, although it is a good electrical conductor, is not a metal, and its resistance behaves in a curious manner: it *decreases* as the temperature is raised. Certain semiconductors behave in the same way; so, incidentally, do insulators. This negative resistance/temperature characteristic is put to good use in 'thermistors'—temperature-sensitive devices which compensate for the rise of resistance of other circuit components.

For example, a thermistor is often included in the heater circuit of radio valves and in the wires supplying current to a film-projector lamp to protect the filament from the very large current which tends to surge through it while it is cold and its resistance is low. Unlike the filament, a cold thermistor has a very high resistance and allows only a small current to pass. However, this small current is enough to heat the thermistor, its temperature slowly rises and its resistance slowly decreases. Thus the current passing through both the thermistor and the delicate tungsten filament increases gradually—there is no sudden surge of current. The resistance of a thermistor can decrease by as much as a factor of 10 million when heated from room temperature to 500° C.

As we saw earlier, the resistance of metals decreases as the temperature goes down. At very low temperatures near the absolute zero (−273° C) the resistance of some metals and alloys suddenly becomes very small indeed, and they are described as *superconducting* (Fig. 9). The resistance of a superconducting circuit is so low that a current continues to flow indefinitely even when the source of e.m.f. is removed. Only the most sophisticated theories of modern physics can account

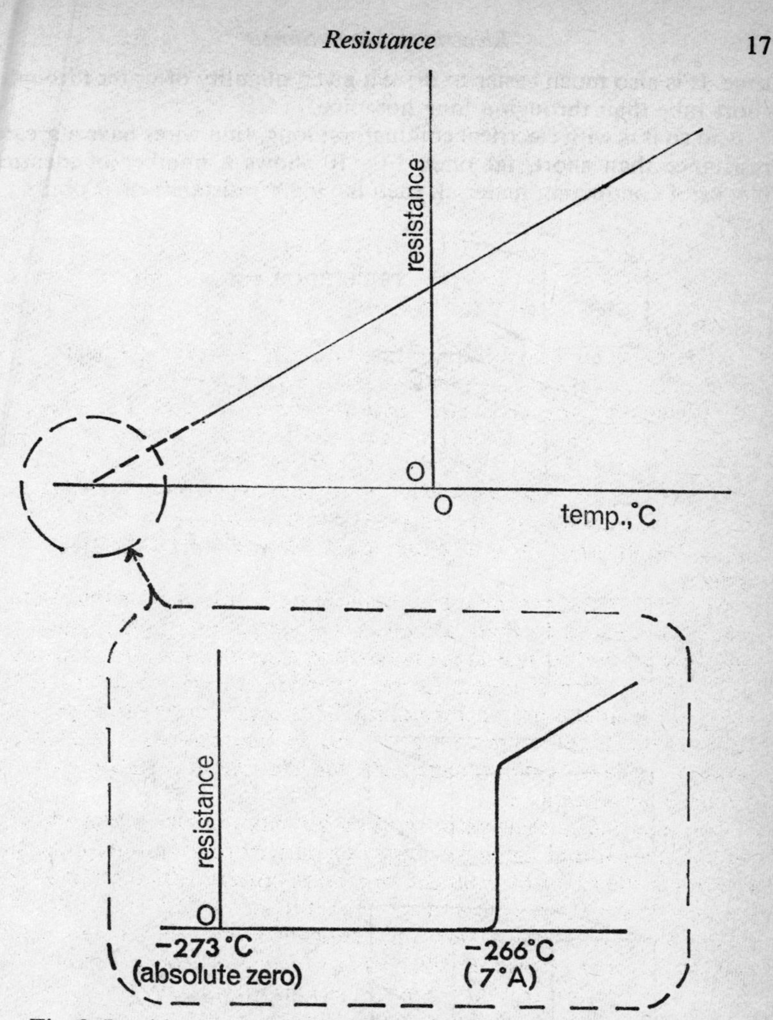

Fig. 9. Lead is a metal which exhibits superconductivity; at 7° from absolute zero its resistance suddenly disappears

for this phenomenon. Our simple picture of electrons as solid particles proves hopelessly inadequate to explain superconductivity.

RESISTIVITY

In Chapter One we used a water-pipe system to represent an electrical circuit. It is fairly clear that the 'resistance' of the water pipes depends on their physical dimensions: it is much easier to drive a given quantity of water through a wide drain-pipe than through a narrow capillary tube. It is also much easier to drive a given quantity of water through a short tube than through a long hosepipe.

And so it is with electrical conductors: long, thin wires have a greater resistance than short, fat ones. Fig. 10 shows a number of identical blocks of conducting material, each having a resistance of R ohms.

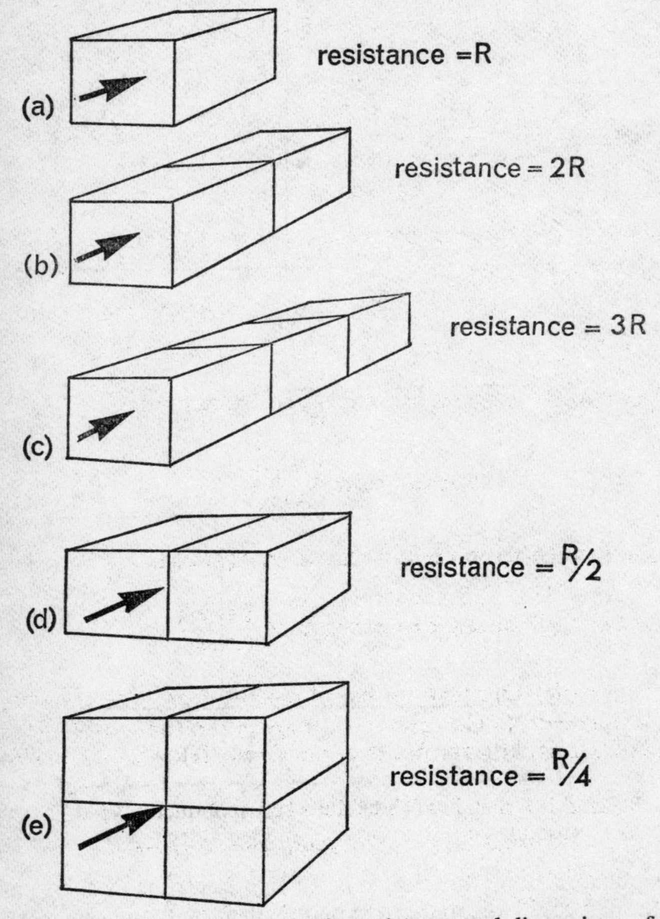

Fig. 10. Relationship between the resistance and dimensions of a conductor

When two blocks are placed end to end the current has to overcome the resistance R of the first and the resistance R of the second: the total resistance to be overcome is plainly $2R$. Similarly, when three blocks are placed end to end the resistance to be overcome is $3R$. A long conductor consisting of a thousand blocks would present a resistance of 1,000R, and so on. Hence the resistance of a conductor is directly proportional to its length.

When two blocks are placed side by side (Fig. 10 (d)), the resistance to be overcome is *not* $2R$, it is only $\frac{1}{2}R$. This is because the two blocks together carry *twice* the current passing through each individual block, and their combined opposition to current is therefore *half* that of a single block. Similarly, the group of four blocks side by side in Fig. 10 (e) together carry four times the current passing through each individual block; their combined resistance is therefore $\frac{1}{4}R$.

A conductor consisting of 1,000 blocks side by side (the actual arrangement makes no difference—its area of cross-section is always 1,000 times that of a single block) would have a combined resistance of $\frac{R}{1{,}000}$. Hence the resistance of a conductor is proportional to $\frac{1}{A}$, where A is its area of cross-section.

The resistance is proportional to the length l of the conductor and proportional to $\frac{1}{A}$; it is therefore (by the rules of mathematics) proportional also to the product of l and $\frac{1}{A}$, so we can write

$$\text{Resistance} \propto \frac{\text{Length}}{\text{Area}}$$

Putting this into the more useful form of an equation,

$$\text{Resistance} = \text{constant} \times \frac{\text{Length}}{\text{Area}}$$

The proportionality constant varies, of course, from one material to another; it is known as the *resistivity* or *specific resistance* of the conducting material, and its symbol is ρ (the Greek letter rho). Hence the

Table 2. RESISTIVITIES OF METALS (at 20° C)

Material	*Resistivity* (Ω*-metres*)
Silver	$1{\cdot}65 \times 10^{-8}$
Copper	$1{\cdot}72 \times 10^{-8}$
Aluminium	$3{\cdot}2 \times 10^{-8}$
Zinc	$5{\cdot}9 \times 10^{-8}$
Nickel	$8{\cdot}7 \times 10^{-8}$
Iron	11×10^{-8}
Manganin	45×10^{-8}
Constantan	49×10^{-8}
Nichrome	112×10^{-8}

formula for the opposition that a conductor offers to the current passing through it is

$$R = \frac{\rho l}{A}$$

where R is measured in ohms, l is the length in metres, A is the area of cross-section in m^2, and ρ is the resistivity in ohm-metres (Ω-m).

The last three materials in Table 2 are nickel alloys frequently used for making resistance wires. Nichrome has a very high resistivity and is made into heating elements for electric fires, etc. Manganin and Constantan have high resistivities combined with low temperature coefficients of resistance (see Table 1): they are employed in standard resistors of high precision.

EXAMPLE: What is the resistance of 100 metres of aluminium wire whose diameter is 2 mm?

From Table 2 the resistivity of aluminium is $3{\cdot}2 \times 10^{-8}$ Ω-m. The length of the wire is 100 m. The area of cross-section has to be found from the formula $A = \pi r^2$, where r is the radius, and $\pi = 3{\cdot}142$. Diameter of the wire is 2 mm, the radius 1 mm = 0·001 m, and the area of cross-section is therefore $3{\cdot}142 \times 0{\cdot}001 \times 0{\cdot}001$ m^2.

$$\begin{aligned}
\rho &= 3{\cdot}2 \times 10^{-8}\,\Omega\text{-m} \\
l &= 100 \text{ m} \\
A &= 3{\cdot}142 \times 0{\cdot}001 \times 0{\cdot}001 \text{ m}^2 \\
R &= \frac{\rho l}{A} \\
&= \frac{3{\cdot}2 \times 10^{-8} \times 100}{3{\cdot}142 \times 0{\cdot}001 \times 0{\cdot}001} \\
&= \frac{3{\cdot}2}{3{\cdot}142} \\
&= 1{\cdot}02\,\Omega \quad \textit{Answer}
\end{aligned}$$

FIXED RESISTORS

All conductors offer some resistance to the electrons flowing through them. Conductors that are deliberately included in a circuit to provide a controlled amount of resistance are called *resistors*.

Resistors are available in a variety of shapes and sizes (some of which are shown in Fig. 11), ranging from small carbon types ¼ in. long used in portable radios to high-power current controllers.

Carbon resistors (Fig. 11 (*a*)) are cylinders of compressed graphite with connecting leads embedded in the ends. They may have any value up to several million ohms (megohms), and the resistance value is usually indicated by a code of brightly coloured bands. They are inexpensive, but should not be used where great precision is required.

Metal-film types (Fig. 11 (*c*)) are made by spraying a thin layer of metal on a glass rod. Because the film is thin, its area of cross-section is small; hence the resistance ($R \propto 1/A$) can be quite high. The film can be etched away in places to adjust its resistance to exactly the value required.

Wire-wound resistors (Fig. 11 (*d*)) are employed for higher powers and greater precision compared with the types already mentioned. The wire used must have a low-temperature coefficient of resistance (see

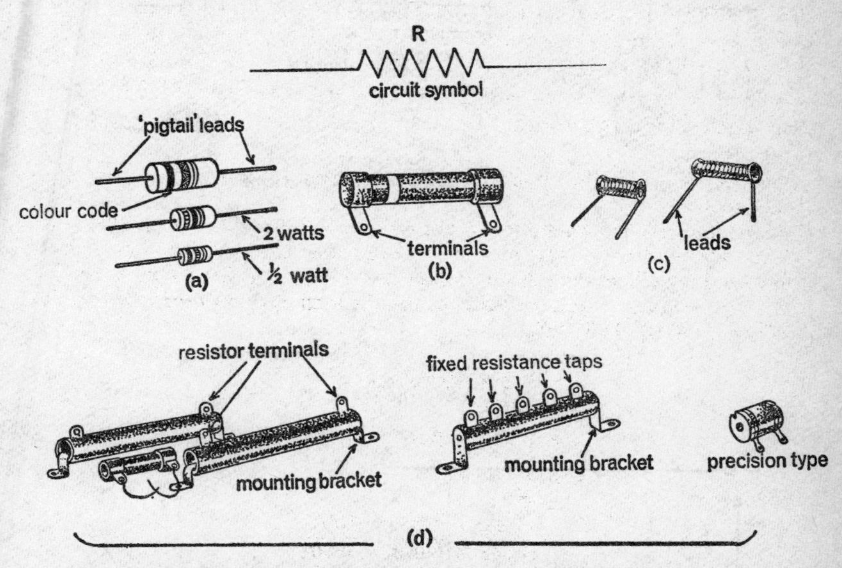

Fig. 11. Fixed resistors. (*a*) Carbon-rod types. (*b*) Larger carbon type. (*c*) Metal-film type. (*d*) Wire-wound types

Table 1), and for this reason the nickel–chromium alloy called Nichrome is the material normally chosen.

VARIABLE RESISTORS

Resistors whose values can be adjusted—like the volume control in a radio receiver—are shown in Fig. 12. The wire-wound stick types (Fig. 12 (*b*)) are used for regulating heavy currents. As the sliding band is moved towards the end of the resistor, the current has to flow through less and less Nichrome wire until, when the sliding band touches the end terminal, the resistance of the stick is nil.

For low-power work, particularly in electronics, it is more convenient to use the rotary type of variable resistor, where the resistance can be increased or decreased steadily by turning a knob. The carbon rotary type (Fig. 12 (*a*)) has a flat ring of graphite, while the wire-wound version (Fig. 12 (*c*)) has a coil of bare resistance wire wound on a ring-shaped former. In both cases the desired value of resistance is tapped off by a rotary contact sweeping over the exposed ring.

A variable resistor in which only two terminals are used—one fixed and one sliding—is called a *rheostat* (from the Greek word meaning

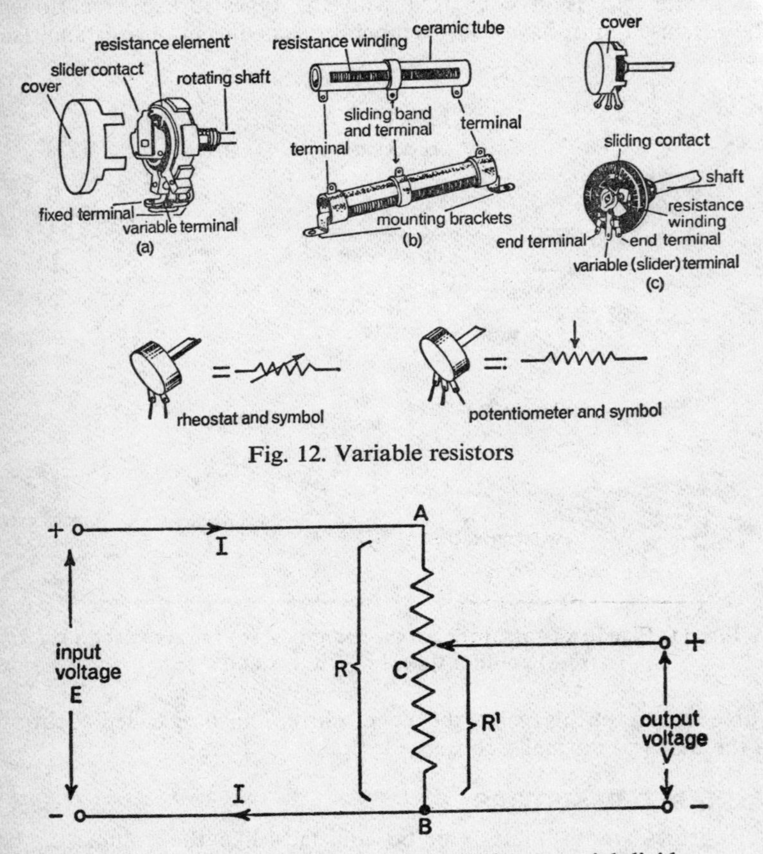

Fig. 12. Variable resistors

Fig. 13. Voltage-divider circuit, also called a potential divider

'flow controller'). If three terminals are used—two fixed and one sliding between them—the variable resistor is known as a *potentiometer*.

The potentiometer provides an adjustable 'voltage-divider' circuit which is a useful means of obtaining various voltages from a fixed potential difference. Fig. 13 shows how the fixed p.d (labelled 'input voltage') is applied to the ends A and B of the potentiometer, while the output voltage is tapped off between the sliding contact C and the fixed end B.

When the sliding contact is moved to A it taps off the whole of the input voltage E. When it is moved to B there is no output voltage at all. At some intermediate point such as C, the output voltage is V, as shown.

Applying Ohm's law to the input circuit:

$$E = IR$$

Applying Ohm's law to the output circuit:

$$V = IR'$$

Dividing the second equation by the first:

$$\frac{V}{E} = \frac{R'}{R}$$

Hence the ratio of the output to input voltages is the ratio of the tapped-off resistance to the total resistance of the potentiometer. The same formula would apply, of course, if the variable resistor were replaced by two fixed resistors between AC and BC.

RESISTORS IN SERIES AND IN PARALLEL

When a number of components are wired up end to end in a continuous chain they are said to be joined *in series*. Fairy lights on a Christmas tree are usually wired in series. If one bulb is removed all the rest go out because the same current flows through every bulb in the chain. The current is the same everywhere throughout a series circuit.

When components are wired up in a kind of ladder arrangement, with all the left-hand ends connected together, and all the right-hand ends connected together, they are said to be joined *in parallel*. Ordinary room lights are wired in parallel across the mains supply. If one bulb is removed the others will not be affected.

Fig. 14 shows three resistors joined in series and in parallel. In the first case, the current has to pass through all three resistors; the total resistance met by the current is obviously the sum of the individual resistances (in this case 11 Ω). Using symbols, the total or combined resistance is

$$R = R_1 + R_2 + R_3 + \text{any number of additional resistances}$$

In the case of resistors joined in parallel, the total or combined resistance R is given by the formula

$$\frac{1}{R} = \frac{1}{R_1} + \frac{1}{R_2} + \frac{1}{R_3} +$$

Applying this formula to the resistors in Fig. 14,

$$\frac{1}{R} = \frac{1}{2} + \frac{1}{3} + \frac{1}{6}$$

$$= \frac{3 + 2 + 1}{6}$$

$$R = \frac{6}{3 + 2 + 1}$$

$$= 1\ \Omega$$

So the total resistance is less than any one of the resistors.

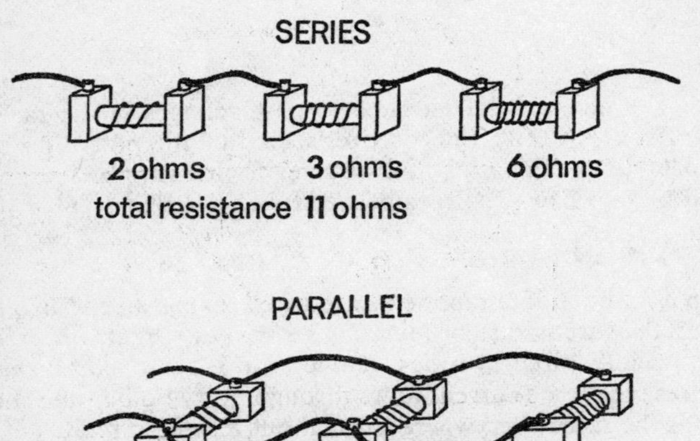

Fig. 14. Resistors joined in series and in parallel

How the formula is obtained for three resistors in parallel is shown in Fig. 15; the same reasoning can be applied to any number of resistors.

The total current flowing is I amps. It divides (at A) into the three currents, I_1, I_2, and I_3 passing through the resistors R_1, R_2, and R_3 respectively. Hence

$$I = I_1 + I_2 + I_3$$

Since all three resistors are connected between points A and B, the p.d. across each is the p.d. between A and B, namely V volts. Applying Ohm's law to each resistor in turn

$$I_1 = \frac{V}{R_1}$$

$$I_2 = \frac{V}{R_2}$$

$$I_3 = \frac{V}{R_3}$$

Applying Ohm's law to the whole network of resistors,

$$I = \frac{V}{R}$$

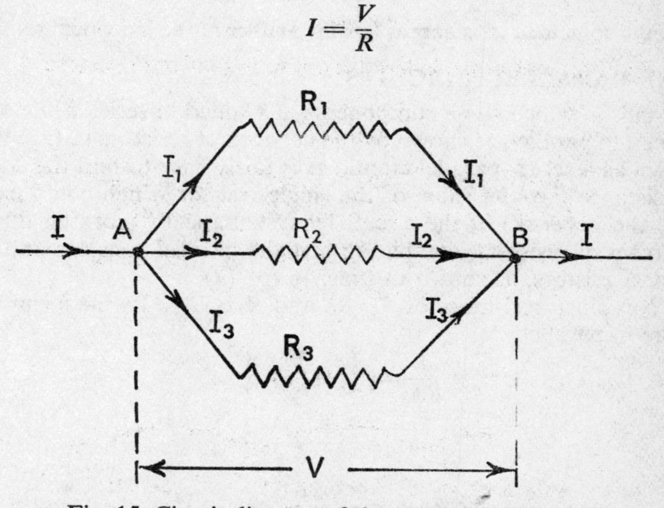

Fig. 15. Circuit diagram of three resistors in parallel

where R is the combined resistance of R_1, R_2, and R_3, i.e. the value of the single resistor which could exactly replace the three. Substituting these relationships, the equation

$$I = I_1 + I_2 + I_3$$

becomes

$$\frac{V}{R} = \frac{V}{R_1} + \frac{V}{R_2} + \frac{V}{R_3}$$

dividing by V

$$\frac{1}{R} = \frac{1}{R_1} + \frac{1}{R_2} + \frac{1}{R_3}$$

EXAMPLE: What is the combined resistance of a 20-Ω resistor in parallel with an 80-Ω resistor?

$$R_1 = 20\ \Omega$$
$$R_2 = 80\ \Omega$$
$$\frac{1}{R} = \frac{1}{R_1} + \frac{1}{R_2}$$
$$\frac{1}{R} = \frac{1}{20} + \frac{1}{80}$$
$$= \tfrac{5}{80}$$
$$R = \tfrac{80}{5}$$
$$= 16\ \Omega \quad \textit{Answer}$$

(Again, the combined resistance is less than either of the individual resistors.)

SERIES–PARALLEL CIRCUITS

A circuit in which some components are joined in series while others are joined in parallel, as shown by the network of resistors in Fig. 16 (*a*), is known as a series–parallel circuit. It is fairly easy to find the equivalent resistance (i.e. the value of the single resistor which could exactly replace the network) of the circuit by systematically working through each group of resistors, combining first the parallel resistors and then the series resistors, as shown in Figs. 16 (*b*)–(*d*).

The combined resistance of R_1, R_2, and R_3 is R_{123}. By the formula for resistors in parallel,

$$\frac{1}{R_{123}} = \frac{1}{3} + \frac{1}{6} + \frac{1}{2}$$
$$= \frac{2 + 1 + 3}{6}$$
$$R_{123} = \frac{6}{2 + 1 + 3}$$
$$= 1\ \Omega$$

The combined resistance of parallel resistors R_6 and R_7 is R_{67}.

$$\frac{1}{R_{67}} = \frac{1}{20} + \frac{1}{5}$$
$$= \frac{1 + 4}{20}$$
$$R_{67} = \frac{20}{1 + 4}$$
$$= 4\ \Omega \quad \text{(Fig. 16 (}b\text{))}$$

The combined resistance of R_{123} and R_4 is $R_{1\text{-}4}$. By the formula for resistors in series,

$$R_{1\text{-}4} = 1 + 0{\cdot}6$$
$$= 1{\cdot}6\ \Omega$$

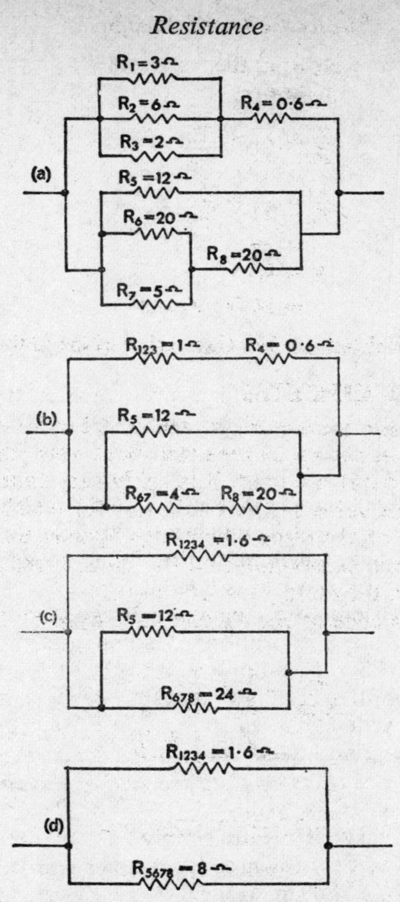

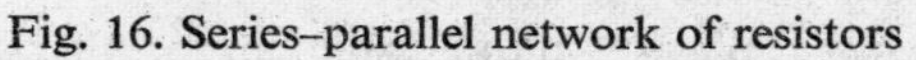
Fig. 16. Series–parallel network of resistors

The combined resistance of the series resistors R_{67} and R_8 is R_{678}

$$R_{678} = 4 + 20$$
$$= 24\ \Omega \quad \text{(Fig. 16 }(c))$$

The combined resistance of parallel resistors R_5 and R_{678} is $R_{5\text{-}8}$

$$\frac{1}{R_{5\text{-}8}} = \frac{1}{12} + \frac{1}{24}$$
$$= \frac{2+1}{24}$$
$$R_{5\text{-}8} = \frac{24}{2+1}$$
$$= 8\ \Omega \quad \text{(Fig. 16 }(d))$$

The combined resistance of parallel resistors $R_{1\text{-}4}$ and $R_{5\text{-}8}$ is $R_{1\text{-}8}$

$$\frac{1}{R_{1\text{-}8}} = \frac{1}{1{\cdot}6} + \frac{1}{8}$$

$$= \frac{10+2}{16}$$

$$R_{1\text{-}8} = \frac{16}{10+2}$$

$$= \frac{4}{3}\,\Omega$$

Thus a single $\frac{4}{3}$-Ω resistor could exactly replace the eight resistors shown in Fig. 16 (*a*).

KIRCHHOFF'S LAWS

Although it is quite a simple matter to find the resistance of a network such as that in Fig. 16 (*a*), it is much more difficult to find the currents flowing in each branch. Currents can be found by applying Ohm's law to each resistor in turn, but this is a tedious process. Fortunately, there are two extremely useful short cuts which can be applied to network analysis. These are generalizations first suggested by the German physicist Gustav Robert Kirchhoff in 1847, and are known as Kirchhoff's laws. Kirchhoff's first law states that *the sum of the currents flowing into any junction is equal to the sum of the currents flowing out of it*. Kirchhoff's second law states that *the total electromotive force in any closed loop of a circle is equal to the sum of the potential differences across the resistances in the loop.*

The first law is easy: it means simply that as many electrons flow away from a point as flow towards it; in other words, there is no accumulation of electrons at a junction.

The second law calls for more explanation. Electromotive force (e.m.f.) is the 'open-circuit voltage' of a battery or generator. The potential difference across each resistance in the loop is, according to Ohm's law, the product *current* × *resistance*. In Fig. 17 (*b*) the 12-volt e.m.f. in loop ABEF is responsible for setting up potential differences between A and B and between B and E. The p.d. between A and B is I_3R_3 volts; that between B and E is I_1R_1 volts. Hence $I_3R_3 + I_1R_1 = 12$. The 10-volt e.m.f. in loop CBED is responsible for setting up potential differences between B and C and between B and E. The p.d. between B and C is I_2R_2 volts; that between B and E is I_1R_1 volts. Hence $I_2R_2 + I_1R_1 = 10$.

When using Kirchhoff's laws for circuit analysis, a systematic procedure has to be followed or the result will be a confusing maze of equations. The recommended procedure is:

(i) Divide the circuit into a number of closed loops.
(ii) Assign a direction to the current flow around each loop.
(iii) Apply the first law to as many junctions as possible.
(iv) Apply the second law to each of the loops.
(v) Solve the resulting simultaneous equations for the required current or currents.

(a)

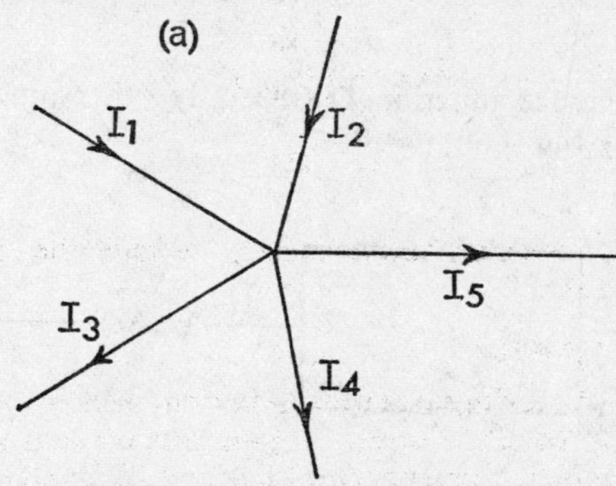

currents flowing into junction $= I_1 + I_2$
currents flowing out of junction $= I_3 + I_4 + I_5$
$I_1 + I_2 = I_3 + I_4 + I_5$

(b)

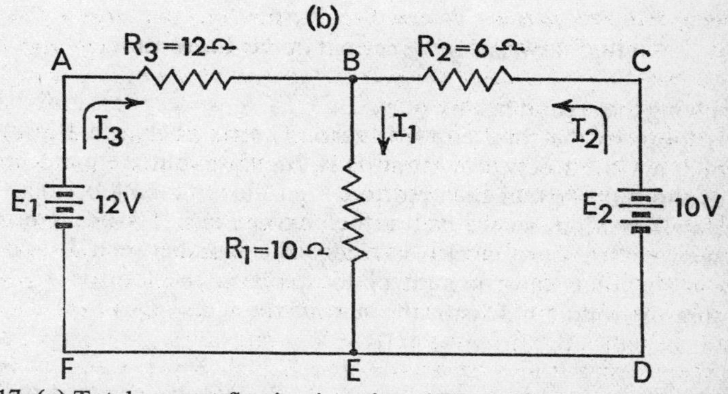

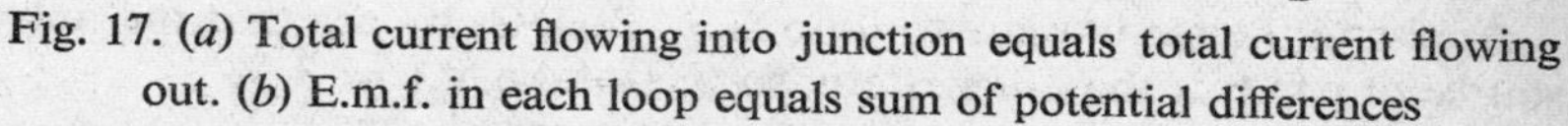
Fig. 17. (*a*) Total current flowing into junction equals total current flowing out. (*b*) E.m.f. in each loop equals sum of potential differences

Fig. 18 shows a circuit whose analysis will illustrate the above procedure. A 2-Ω resistor and a 4-Ω resistor are connected in parallel, and the combination is joined in series with a 5-Ω resistor and a 3-V battery whose internal resistance is 0·8 Ω. What is the current in the 4-Ω resistor?

The circuit contains two obvious closed loops, ABCDEFA and GHABCDG. Arrows have been drawn to show the assumed direction of the current in each branch of the circuit; if our assumptions are wrong the currents we calculate will still be correct in size, but they will have minus signs in front of them.

Applying the first law to junction A gives

$$I_1 = I_2 + I_3 \quad (1)$$

Applying the first law to junction D (the only other junction in this circuit) gives exactly the same result.

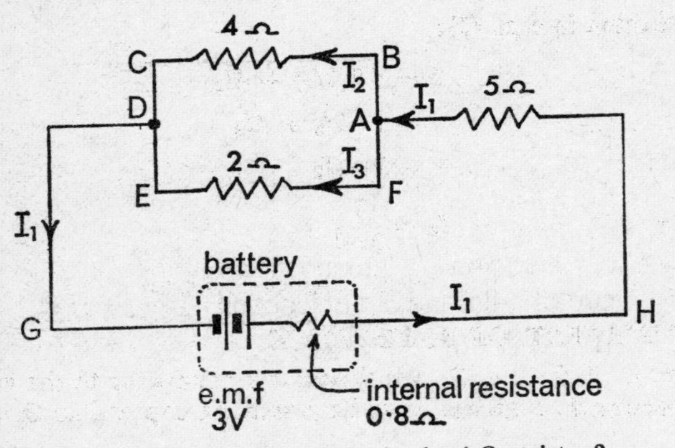

Fig. 18. What is the current in the 4-Ω resistor?

Applying the second law to loop ABCDEFA, we see there is no battery in this loop, so the total e.m.f. is zero. The p.d. between B and C is $4I_2$ volts, and that between E and F is $2I_3$ volts, but we must be very careful about the *sign* of each p.d. We are following the loop in an anti-clockwise direction, so we call anti-clockwise currents positive, while clockwise currents are negative. Hence the p.d. between E and F is really $-2I_3$ volts, and the sum of the p.d.s in the loop is $4I_2 - 2I_3$. Equating the total e.m.f. with the sum of the p.d.s gives

$$0 = 4I_2 - 2I_3 \quad (2)$$

Applying the second law to loop GHABCDG, we see that the only source of e.m.f. is the 3-V battery, so the total e.m.f. = 3 V.

The p.d. between G and H is $0{\cdot}8I_1$ volts, that between H and A is $5I_1$ volts and that between B and C is $4I_2$ volts. (All these are positive, since the currents are anti-clockwise.) Sum of the p.d.s in this loop is $0{\cdot}8I_1 + 5I_1 + 4I_2$. Equating the total e.m.f. with the sum of the p.d.s gives

$$3 = 5{\cdot}8I_1 + 4I_2 \quad (3)$$

We now have three simultaneous equations which can be solved to find I_2

From eqn. (2):

$$2I_3 = 4I_2$$

hence $$I_3 = 2I_2$$

Substituting in eqn. (1):

$$I_1 = I_2 + 2I_2$$

hence $$I_1 = 3I_2$$

Substituting in eqn. (3):

$$\begin{aligned} 3 &= 5{\cdot}8(3I_2) + 4I_2 \\ &= 17{\cdot}4I_2 + 4I_2 \\ &= 21{\cdot}4I_2 \end{aligned}$$

Hence $$\begin{aligned} I_2 &= \frac{3}{21{\cdot}4} \\ &= 0{\cdot}14 \text{ amps} \end{aligned}$$ *Answer*

MEASUREMENT OF RESISTANCE

The ratio of the p.d. (in volts) across a conductor to the current (in amps) passing through it is, as we have seen in Chapter One, the *resistance*

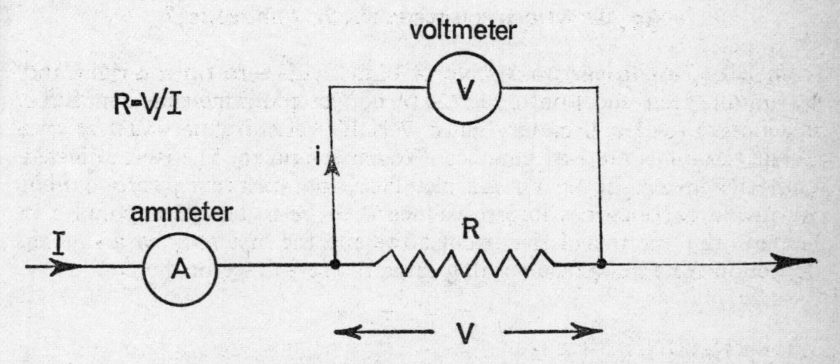

Fig. 19. Measurement of resistance with voltmeter and ammeter

(in ohms) of the conductor. So if we place a voltmeter across the conductor to measure the p.d. and an ammeter in series to measure the current, as shown in Fig. 19, the ratio of the two meter readings gives the resistance of the conductor.

The snag with this simple scheme, apart from the fact that it calls for an ammeter and a voltmeter of high quality, is that some of the current by-passes the conductor whose resistance is being measured. This is the small current (labelled i in Fig. 19) which flows through the voltmeter. Hence the current through the conductor is not I (the current through the ammeter) but $I - i$, and the ratio V/I is only approximately equal to the resistance being measured.

For a quick measurement of resistance, where accuracy is a secondary consideration, an instrument called an *ohmmeter* may be used (Fig. 20). This kind of instrument often forms part of a 'multimeter' which, at the turn of a switch, measures current or voltage or resistance. Basically, it is an ammeter which, instead of having its scale graduated in amperes

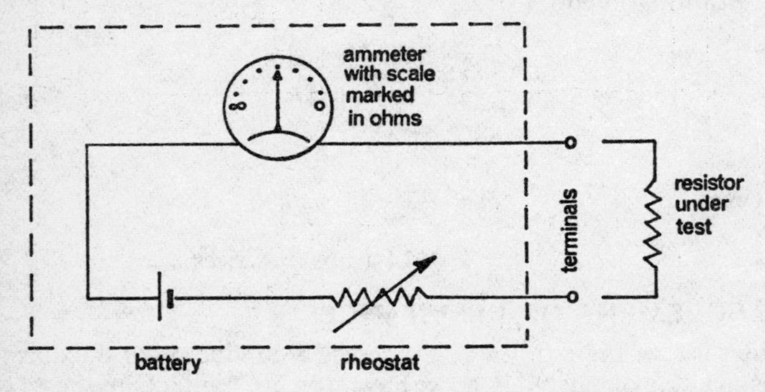

Fig. 20. Ohmmeter circuit; the resistance of any conductor connected across the terminals is read directly from the scale

from left to right, carries a scale of ohms with zero on the right and ∞ (infinite resistance) on the left. The converted ammeter is connected in series with a small battery and a variable resistor (rheostat).

The terminals are first connected together and the rheostat adjusted until the meter shows a full-scale deflection (i.e. reads zero ohms). When the conductor whose resistance is to be measured is connected between the terminals less current flows and the meter shows a smaller deflection. The new scale reading gives the resistance of the conductor directly.

WHEATSTONE BRIDGE

For resistance measurements of high precision, standard resistors are available whose values are accurately known. Foremost among the methods of comparing an unknown resistance with a standard resistor is that devised by Sir Charles Wheatstone (as long ago as 1833) and known as Wheatstone's bridge.

The apparatus shown in Fig. 21 (*a*) is a students' form of Wheatstone bridge: laboratory versions are, of course, more sophisticated than this,

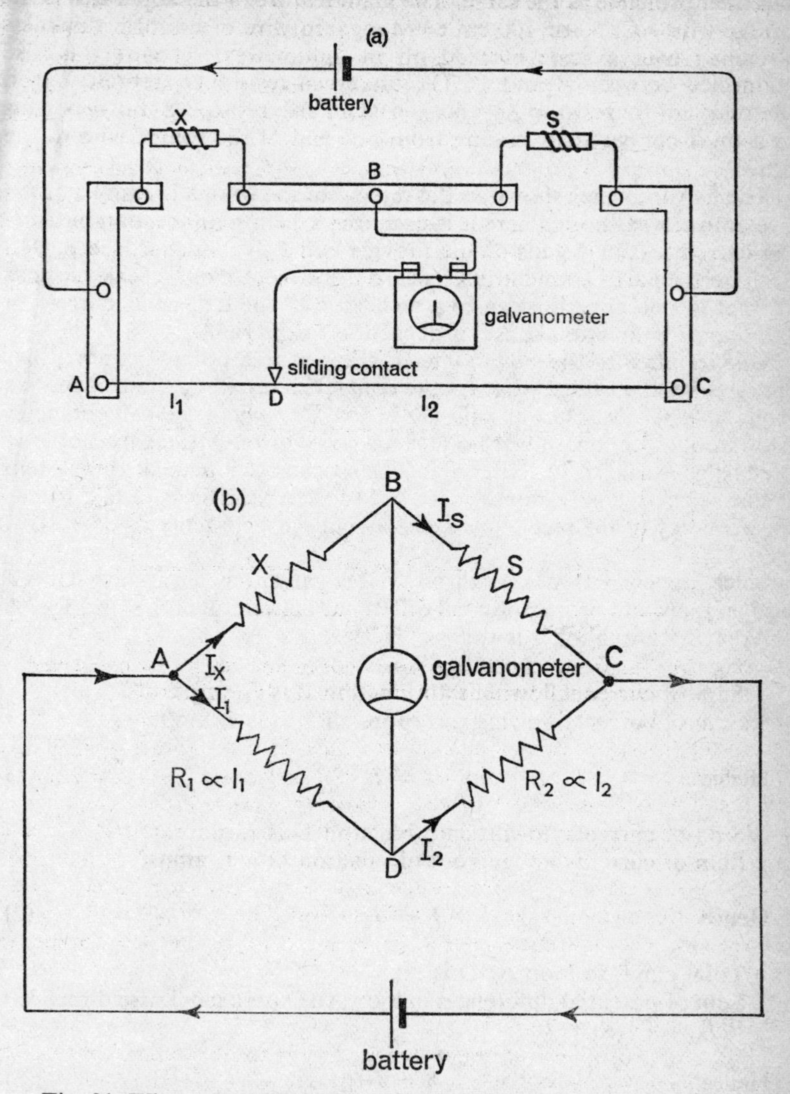

Fig. 21. Wheatstone's bridge. (*a*) Layout of components. (*b*) Circuit diagram

but their principle is the same. The main feature of the apparatus is the bridge wire AC, 50 or 100 cm of Manganin wire of uniform thickness stretched over a scale marked off in millimetres. A battery is also connected between A and C. The unknown resistance (represented in the diagram by resistor X) and the standard resistor S fill two gaps in a thick copper strip leading from one end of the bridge wire to the other.

The galvanometer shown in the middle of the bridge is really a sensitive ammeter, although here it is used for detecting rather than measuring currents. (For details of the moving-coil galvanometer, see p. 92.) This particular instrument has its zero in the centre of its scale, so that current in one direction causes a deflection to the left, while current in the opposite direction causes a deflection to the right.

One terminal of the galvanometer is joined to a pointed contact that slides along the bridge wire. As the contact slides along the bridge wire from A to C it reaches a point (D in Fig. 21) where the galvanometer shows no deflection. When no current flows through the galvanometer the bridge is said to be 'balanced'. The special advantage of this system is that, since the galvanometer is used to detect the *absence* of current, the accuracy of the result is less dependent on the quality of the instrument.

Since no current passes through the galvanometer in a balanced bridge, there can be no potential difference between B and D in Fig. 21.

Applying Kirchhoff's laws:

Sum of currents flowing into junction B is I_x amps.
Sum of currents flowing out of junction B is I_s amps.

Hence $$I_x = I_s \qquad (1)$$

Sum of currents flowing into junction D is I_1 amps.
Sum of currents flowing out of junction D is I_2 amps.

Hence $$I_1 = I_2 \qquad (2)$$

Total e.m.f. in loop ABD is nil.
Sum of potential differences in loop ABD (in a clockwise direction) is $I_xX - I_1R_1 = 0$

Hence $$I_xX = I_1R_1 \qquad (3)$$

where R_1 is the resistance of the bridge wire between A and D.
Total e.m.f. in loop BCD is nil.
Sum of potential differences in loop BCD (in a clockwise direction) is $I_sS - I_2R_2 = 0$

Hence $$I_sS = I_2R_2 \tag{4}$$

where R_2 is the resistance of the bridge wire between C and D.

Dividing eqn. (3) by eqn. (4),

$$\frac{I_xX}{I_sS} = \frac{I_1R_1}{I_2R_2}$$

but $$I_x = I_s \quad \text{and} \quad I_1 = I_2$$

Hence $$\frac{X}{S} = \frac{R_1}{R_2} \tag{5}$$

If the values of R_1, R_2 and S are known the unknown resistance X can be found from eqn. (5).

However, there is no need to know the actual resistance R_1 between A and D or the resistance R_2 between C and D. The resistance between A and D is, according to the formula given on p. 20,

$$R_1 = \frac{\rho l_1}{A}$$

where ρ is the resistivity of Manganin, l_1 is the length of bridge wire between A and D, and A is the cross-sectional area of the bridge wire. Similarly,

$$R_2 = \frac{\rho l_2}{A}$$

where l_2 is the length of wire between C and D.

Hence $$\frac{R_1}{R_2} = \frac{l_1}{l_2}$$

and eqn. (5) can be written as

$$\frac{X}{S} = \frac{l_1}{l_2}$$

or $$X = S\frac{l_1}{l_2} \tag{6}$$

Since the value of the standard resistor S is accurately known, the unknown resistance X can be found by reading off the lengths l_1 and l_2 from the scale under the bridge wire and using the result in eqn. (6).

EXAMPLE: A platinum resistance thermometer (see p. 16) is connected in one arm of a Wheatstone bridge, and a standard 10-Ω resistor is connected in the other. The bridge is balanced when the sliding contact is 125 mm from the end connected to the thermometer. If the bridge wire is 0·5 m long, what is the resistance of the thermometer?

$$l_1 = 125 \text{ mm}$$
$$l_2 = 500 - 125$$
$$= 375 \text{ mm}$$
$$S = 10\ \Omega$$

Using eqn. (6), the resistance of the thermometer is

$$X = S\frac{l_1}{l_2}$$
$$= 10 \times \frac{125}{375}$$
$$= 3\tfrac{1}{3}\ \Omega \quad \textit{Answer}$$

SUMMARY

When a battery or generator delivers a current I, the p.d. between its terminals V is less than the open-circuit voltage (e.m.f.) E by an amount equal to the p.d. across the *internal resistance* r. Thus $V = E - Ir$.

The resistance of metallic conductors increases with temperature; the relationship between resistance and temperature is given by the equation $R_t = R_0(1 + \alpha t)$, where α stands for the temperature coefficient of resistance of the metal.

The resistance of carbon and certain semiconductors decreases as the temperature increases.

Materials which have practically no resistance at extremely low temperatures are described as superconducting.

The resistance of a conductor is directly proportional to its length and inversely proportional to its cross-sectional area. The relationship is given by the formula

$$R = \rho\frac{l}{A}$$

where ρ is the *resistivity* of the conducting material.

The combined resistance R of any number of resistors R_1, R_2, R_3, etc., in *series* is given by $R = R_1 + R_2 + R_3 + \ldots$

The combined resistance R of any number of resistors R_1, R_2, R_3, etc., in *parallel* is given by $\frac{1}{R} = \frac{1}{R_1} + \frac{1}{R_2} + \frac{1}{R_3} + \ldots$

Kirchhoff's first law states that the sum of the currents flowing into any junction in a circuit is equal to the sum of the currents flowing out of it.

Kirchhoff's second law states that the total e.m.f. in any closed loop of a circuit is equal to the sum of the potential differences across the resistances in the loop.

Resistance can be measured accurately with a Wheatstone bridge. The condition for a balanced bridge is

$$\frac{X}{S} = \frac{l_1}{l_2}$$

where l_1 is the length of bridge wire between the sliding contact and the (unknown) resistance whose value is X, l_2 is the length of bridge wire between the sliding contact and the (standard) resistor whose value is S.

CHAPTER THREE

ELECTROHEAT

Even the best electrical conductors have some resistance, and the electrons drifting through them collide with atoms: the electrons lose energy and the atoms gain energy. As a result, the atoms vibrate more vigorously and the temperature of the conductor tends to rise. In short, electrical energy is converted into heat.

Very often the quantity of heat generated is small enough to escape—by radiation, convection, and thermal conduction—as fast as it is produced; the temperature then remains steady. In some conductors, such as the filament of a lamp or the heating element of an electric blanket, heat is produced at such a rate that there is a considerable rise in temperature.

Electrical appliances which utilize the heating effect of a current, e.g. radiators, kettles, blankets, and light bulbs, are rated in watts and kilowatts. These units give a measure of the *rate* at which electrical energy is consumed by an appliance: they are units of electrical *power*. (They are not units of energy: electrical energy is measured in *joules*.) One watt (1 W) is the power of an appliance which accepts one joule of electrical energy per second and converts it to some other useful form—light, heat, or motive power. One kilowatt (1 kW) is 1,000 watts, and is roughly equal to $1\frac{1}{3}$ horsepower.

There is a very simple relationship between the power rating P of an appliance and the current I which passes through it when the p.d. across its ends is V:

$$P \text{ (watts)} = I \text{ (amps)} \times V \text{ (volts)} \qquad (1)$$

This is a consequence of the way in which the volt is defined (see p. 90).

Using Ohm's law, we can write $I = \frac{V}{R}$, and the power equation becomes

$$P = \frac{V}{R} \times V$$

$$= \frac{V^2}{R} \qquad (2)$$

or we can write $V = IR$, and the power equation becomes

$$P = I \times IR$$

$$= I^2R \qquad (3)$$

EXAMPLE: What current does a 2-kW heater draw from a 250-V supply? What is its resistance?

$$P = 2{,}000 \text{ W}$$
$$V = 250 \text{ V}$$
$$P = IV$$
$$2{,}000 = I \times 250$$
$$I = 8 \text{ amps} \quad \textit{Answer}$$

$$P = \frac{V^2}{R}$$
$$2{,}000 = \frac{250 \times 250}{R}$$
$$R = \frac{250 \times 250}{2{,}000}$$
$$= 31{\cdot}25 \text{ ohms} \quad \textit{Answer}$$

(*Note:* this is the resistance of the heater when hot; initially its resistance will be less than 31·25 Ω and it will draw much more than 8 A from the mains.)

JOULE'S LAW

Power was defined at the beginning of the chapter as the *rate* at which energy is converted from one form to another, i.e.

$$\text{Power} = \frac{\text{Energy}}{\text{Time}}$$

Turning round this definition we see that

$$\text{Energy (joules)} = \text{Power (watts)} \times \text{Time (seconds)}$$

Using the symbol W for energy (the same symbol is used for work) and the symbol t for time

$$W = Pt$$

and we can substitute any of the three power formulae (eqns. (1)–(3)) for P.

Hence the number of joules of electrical energy converted by an appliance is

$$W = VIt \qquad (4)$$

$$W = \frac{V^2 t}{R} \qquad (5)$$

or

$$W = I^2 Rt \qquad (6)$$

where I is current, V is potential difference, and R is resistance.

Eqn. (6) was discovered experimentally in 1841 by James Prescott Joule, an amateur scientist working in Manchester, and is known as Joule's law.

A modern version of the apparatus used by Joule in his experiment is shown in Fig. 22. It consists of a small heating coil connected to a battery and a rheostat. The coil is immersed in water in a well-lagged copper can called a *calorimeter*. A thermometer records the rise in temperature of the water. Knowing the temperature rise, the mass of the water, and the mass of the calorimeter, it is possible to calculate the

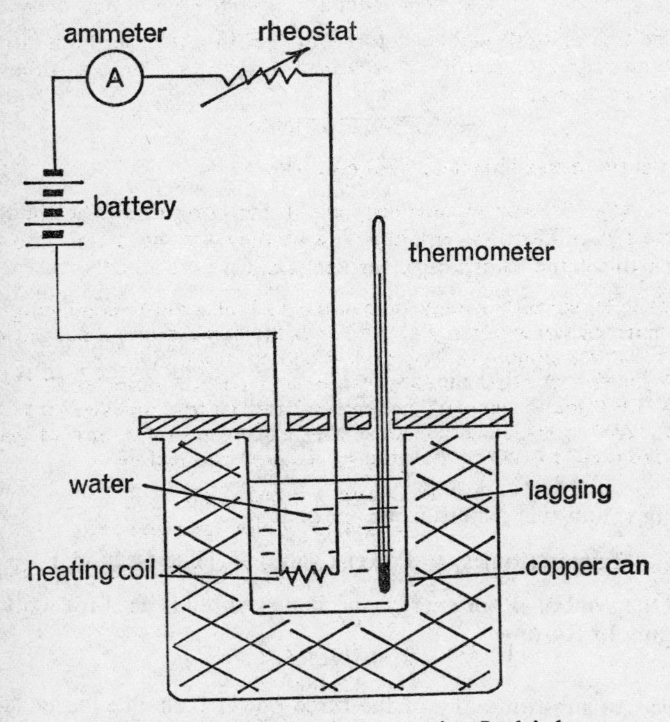

Fig 22. Apparatus for demonstrating Joule's law

quantity of heat energy W given out by the coil, since we know that 4·18 joules will raise the temperature of 1 gram of water by 1° C.

By carrying out the experiment a number of times, as Joule did, it is found that W is proportional to the time t for which the current flows. Using the rheostat to vary the current I amps, it is found that W is proportional to I^2, i.e. to the *square* of the current. (Thus, doubling the current gives *four* times as much heat, trebling the current gives *nine* times as much heat, and so on.) Changing the heating coil for others of different resistance, it is found that W is proportional to the resistance R of the coil.

Mathematically, since W is proportional to the square of the current and to the resistance and to the time, it is also proportional to their product. Hence

$$W \propto I^2Rt$$

or, putting this proportionality into the form of an equation,

$$W = \text{constant} \times I^2Rt$$

If W is measured in joules, I in amps, R in ohms, and t in seconds, the value of the proportionality constant is unity. The equation above therefore becomes

$$W = I^2Rt \text{ joules}$$

which is of course identical with eqn. (6).

Eqns. (4), (5), and (6) are very useful for computing the theoretical heat output of electrical appliance, and they are valid for both direct current (d.c.) and alternating current (a.c.).

EXAMPLE: How much energy is required to boil a kettle containing 1 litre (1·76 pints) of water initially at 20° C? If the kettle draws a current of 4 A from a 240 V supply, how long does it take to boil?

4·18 joules will raise the temperature of 1 cm³ of water by 1° C. Hence 4·18 × 1,000 joules will raise the temperature of 1 litre of water by 1° C, and 4·18 × 1,000 × 80 joules will raise the temperature of 1 litre of water by 80° C from 20° to 100° C. The energy required is therefore

$$\begin{aligned} W &= 4{\cdot}18 \times 1{,}000 \times 80 \\ &= 334{,}400 \text{ joules} \quad \textit{Answer} \end{aligned}$$

Given that $V = 240$ V and $I = 4$ A, we can substitute these values in eqn. (4):

$$\begin{aligned} W &= VIt \\ 334{,}400 &= 240 \times 4 \times t \\ t &= \frac{334{,}400}{240 \times 4} \\ &= 348 \text{ seconds} \\ &= 5{\cdot}8 \text{ minutes} \quad \textit{Answer} \end{aligned}$$

EXAMPLE: An electric oven operated from a 240-V supply has a heating element comprising two 30-Ω coils. The coils may be connected in series to give a 'low' setting, or in parallel to give a 'high' setting. What is the heat output in 10 min for each setting?

With the coils in series, the total resistance is 30 + 30 = 60 Ω

$$\begin{aligned} 10 \text{ min} &= 10 \times 60 = 600 \text{ sec} \\ V &= 240 \text{ V} \\ R &= 60\ \Omega \\ t &= 600 \text{ sec} \end{aligned}$$

Using eqn. (5)

$$W = \frac{V^2t}{R} \text{ joules}$$
$$= \frac{240 \times 240 \times 600}{60}$$
$$= 576{,}000 \text{ joules} \quad \textit{Answer}$$

With the coils in parallel, the total resistance is 15 Ω $\left(\text{from } \frac{1}{R} = \frac{1}{30} + \frac{1}{30}\right)$

$$V = 240 \text{ volts}$$
$$R = 15\ \Omega$$
$$t = 600 \text{ sec}$$
$$W = \frac{V^2t}{R}$$
$$= \frac{240 \times 240 \times 600}{15}$$
$$= 2{,}304{,}000 \text{ joules} \quad \textit{Answer}$$

(Thus the parallel connexion produces four times as much heat as the series connexion.)

ELECTRICITY BILLS

An electrical supply is a source of energy, and naturally enough we pay for the number of 'units' of energy used—not the number of electrons. The usual unit of electrical energy, the joule, is too small for such purposes as assessing the quarterly bill, and the unit used in practice is the kilowatt-hour.

One kilowatt-hour (1 kWh) is the energy expended (or the work done) by a power of 1,000 watts operating for one hour.

$$1 \text{ kWh} = 1{,}000 \text{ W} \times 3{,}600 \text{ sec}$$
$$= 3{,}600{,}000 \text{ W-sec}$$
$$= 3{\cdot}6 \text{ million joules}$$

(which should give some idea why the joule is not used in computing electricity bills).

EXAMPLE: What is the cost of running a 60-W lamp 5 hours every day for a year, if the charge for electricity is one (new) penny per kWh?

60 W = 0·06 kW

5 hours per day for 365 days is 1,825 hours

Therefore the energy consumed is 0·06 × 1,825 kWh, and the cost is

$$0{\cdot}06 \times 1{,}825 \times 1 \text{ penny}$$
$$= £1{\cdot}09\tfrac{1}{2} \quad \textit{Answer}$$

EXAMPLE: A 9-V transistor battery costs $12\frac{1}{2}$ p and delivers a current of 30 mA for 20 hours before it has to be thrown away. What is the cost of electricity per kWh in this case?

$$\begin{aligned}\text{Power} &= I\,(\text{amps}) \times V\,(\text{volts})\\ &= 0{\cdot}03 \times 9\\ &= 0{\cdot}27\ \text{W}\\ &= 0{\cdot}00027\ \text{kW}\end{aligned}$$

A power of 0·00027 kW operating for 20 hours dissipates an energy of

$$0{\cdot}00027 \times 20 = 0{\cdot}0054\ \text{kWh}$$

If 0·0054 kWh of energy costs $12\frac{1}{2}$ p

$$\begin{aligned}1\ \text{kWh of energy costs}\ &\frac{12\frac{1}{2}}{0{\cdot}0054}\ \text{p}\\ &= £23{\cdot}15 \quad \textit{Answer}\end{aligned}$$

KILOWATT-HOUR METER

The number of kilowatt-hours of electrical energy taken from the mains supply is measured by the consumer's 'electricity meter' (Fig. 23). The meter is in fact a small electric motor which turns a series of pointers moving over dials marked off in ten-thousands, thousands, hundreds, tens, units, and tenths of kilowatt-hours.

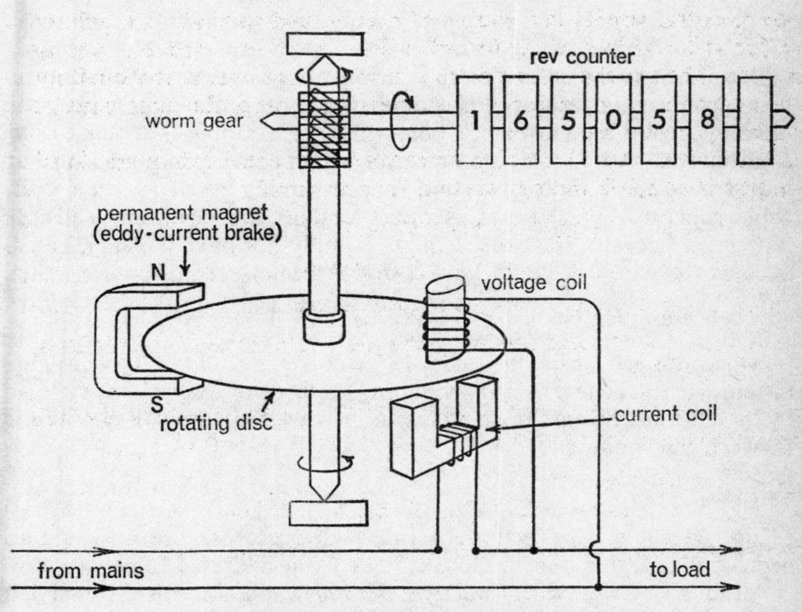

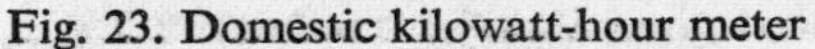
Fig. 23. Domestic kilowatt-hour meter

As well as the dials, the front of the meter usually has a small window through which an aluminium disc can be seen rotating. Inside the meter there is a pair of fixed coils, one below the disc and one above it.

Nearly all the current drawn from the mains passes through the lower coil; a small fraction of it flows through the upper coil, and this fraction is proportional to the mains voltage. The *speed* at which the disc rotates is proportional to the current in the lower coil and to the current in the upper coil. Since the latter current is proportional to the voltage, the speed of rotation is proportional to both the mains current and the mains voltage. Thus the greater the *power* (current × voltage) drawn from the mains, the faster the disc rotates.

The energy supplied by the mains is power × time, hence the speed of rotation × time is proportional to the energy being measured. But the speed of rotation × time is plainly the *total* number of rotations made in a given period. (E.g. if the disc rotates at a speed of 15 rev/min, in 10 minutes it makes a total of 15 × 10 = 150 revolutions.) Hence counting the total number of revolutions gives a measure of the energy supplied, and this is just what the meter does.

HEATING EFFECT OF CURRENT

Joule's law states that the heat energy produced by a current I flowing for t sec in a conductor of resistance R ohms, is given by

$$W = I^2Rt$$

Because it is the *square* of the current that governs the quantity of heat produced, overheating becomes a major problem in apparatus handling really big currents. Comparing, for example, a wire carrying a current of 1 A to a television receiver with a cable carrying 60 A to an industrial electric motor, we find that, assuming equal resistance, the cable generates 3,600 times as much heat as the television leads. In practice, of course, the cable would certainly not have the same resistance as the wire—it would have a much greater area of cross-section, and hence a much smaller resistance $\left(R = \frac{\rho l}{A}\right)$.

Apparatus which has to carry heavy currents is carefully ventilated to remove heat as fast as it is produced. Motors, for example, are cooled by blowing air through them; transformers are cooled by oil circulating between the windings.

FUSES

Fig. 24 shows a selection of fuses—devices that are designed to protect electrical appliances and circuits from excessive currents.

Although rarely used nowadays, the renewable link fuse is included in Fig. 24 because it illustrates the basic action of all fuses. It is a thin strip of copper or lead alloy having a high resistivity and a low melting point compared with the materials normally used for making conductors. The fuse link, clamped in an appropriate holder, is wired in series with the circuit that it protects so that the same current flows through both.

For normal currents the heat developed in the fuse link escapes to the surroundings as quickly as it is produced. But if the current increases, the amount of heat developed—which is proportional to the *square* of the current—is very much greater and the temperature rises until the alloy strip melts. The word fuse literally means 'to melt'. In this instance the melting takes place at the two constrictions; the central portion drops out, leaving a gap too wide for the current to spark across.

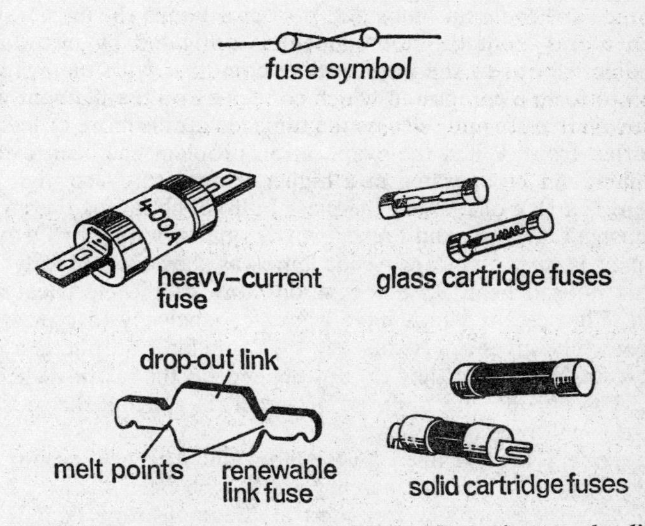

Fig. 24. Types of fuse used to protect circuits against overloading

Glass cartridge fuses are used mainly in the protection of electronics apparatus, where currents are normally small. They contain a thin wire of low-melting-point alloy stretched between metal caps at the end of a glass tube. 'Solid' cartridge fuses used with domestic appliances are similar, except that the alloy wire is enclosed in a porcelain tube. The tube is a guard against the risk of fire when the fuse melts; the thin wire breaks very rapidly indeed and there is a small-scale explosion. In up-to-date house wiring a solid cartridge fuse is normally contained in the plug attached to each appliance. Such fuses are marked with the size of the steady current (2 A, 5 A, 10 A, or 13 A) that they will safely allow to pass.

APPLIANCES

The majority of domestic electrical appliances depend on the heating effect of a current. Even radio and television receivers use the effect in their thermionic valves: the electrons flowing through each valve are emitted from a cathode which is kept at a high temperature by the heat produced in a tungsten filament.

Light bulbs also contain a thin tungsten filament, usually in an atmosphere of argon and nitrogen. Current passing through the filament raises its temperature to about 2,300° C, at which temperature it is white hot and emits light—the hotter the filament, the whiter the light. Unfortunately, white-hot metals emit atoms as well as light, so that the filament gets thinner as it ages, until it eventually breaks.

In some car-headlamp bulbs and projector lamps the evaporation of tungsten atoms from the hot filament is mitigated by introducing a little iodine vapour to the bulb. Iodine combines with the evaporated tungsten to form a compound which condenses on the filament when it cools down after use and releases the tungsten atoms more or less where they started from. When the evaporation problem has been overcome the filament can be operated at a higher temperature—so high in fact that there is a risk of melting the glass bulb. Such lamps therefore use quartz instead of glass, and are known as quartz–iodine or QI types.

Filament lamps, or incandescent lamps as they are properly called, waste, as unused heat, 95 per cent or more of the electrical energy supplied. Fluorescent lamps have a greater efficiency (a typical 40-W fluorescent tube gives approximately the same light output as a 100-W incandescent lamp), but they do not depend on the heating effect of a current. The heating effect is used, however, in starting the passage of current through the lamp.

A fluorescent tube is filled with argon and mercury vapour under reduced pressure. When the mercury atoms lose electrons they become positively charged mercury ions and at the same time emit invisible ultraviolet radiation. It is the ions which conduct the current through the tube—there is no metallic conductor present.

The process of ionization has to be started by means of a very high voltage (produced in a coil called a 'ballast', see p. 116) combined with a supply of electrons from a hot filament at each end of the tube. Once ions are present, the normal mains voltage is sufficient to drive current through the tube. The light in a fluorescent lamp comes from the 'phosphors' with which it is coated: these give out white light when the ultraviolet radiation produced by ionization falls on them.

Electric fires have heating elements made from Nichrome, or one of the other nickel alloys, wound on a fireclay support. A convector heater uses the same type of coiled heating element, but operates at a rather lower temperature. Kettles, irons, immersion heaters, etc., have their

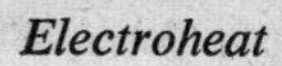

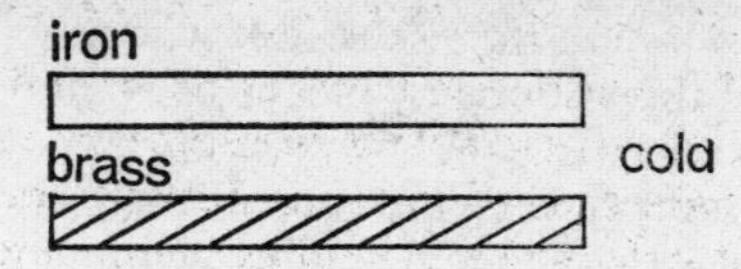

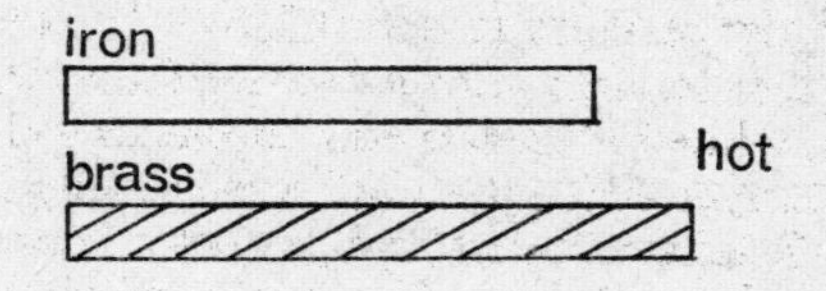

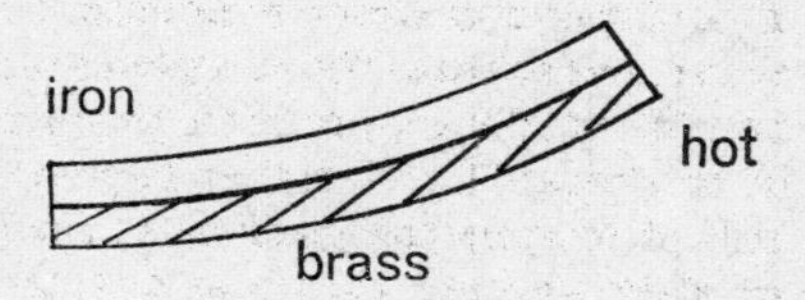

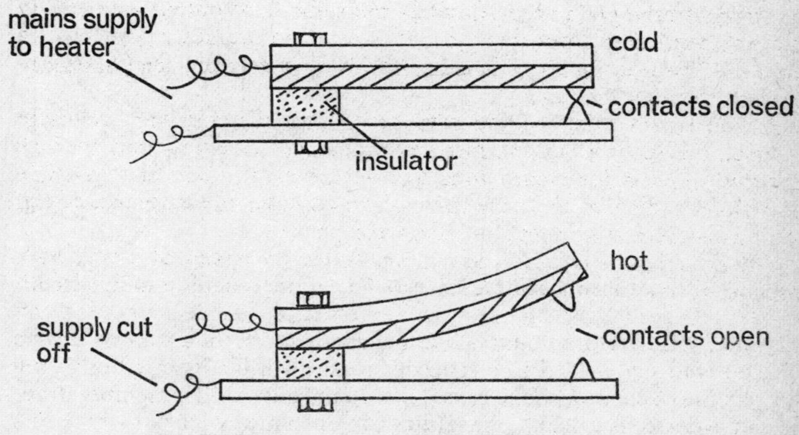

Fig. 25. Principle of the bimetallic-strip thermostat

heating elements totally enclosed. The two last-mentioned appliances are usually able to switch themselves off when the required temperature has been reached; this is done by means of an automatic switch called a *thermostat*.

The most common type of thermostat used in domestic applications employs the principle of the bimetallic strip shown in Fig. 25. When equal lengths of different metals, such as iron and brass, are heated they

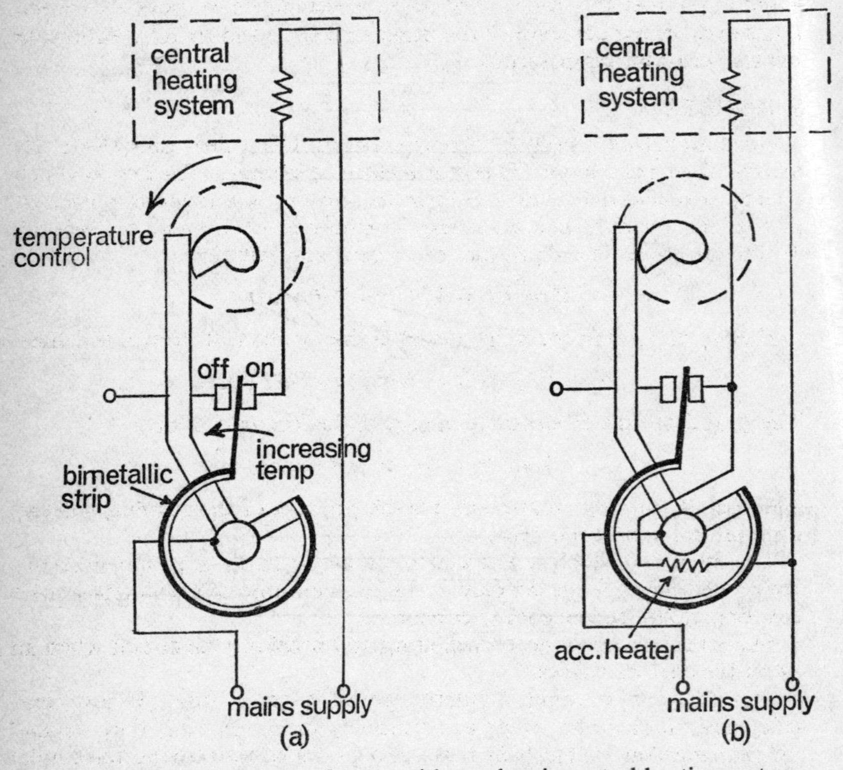

Fig. 26. Room thermostat for use with an electric central-heating system

expand by different amounts: brass expands more than iron. If a strip of iron is firmly welded to a strip of brass it will be distorted on being heated, into a curve with the brass strip on the outside. This temperature-sensitive device is called a bimetallic strip or 'bimetal'. By attaching one end of a bimetallic strip to one of the contacts of a switch the bending of the strip can be arranged to break a circuit as the temperature rises. When the temperature falls, the strip straightens out again and closes the circuit.

A practicable thermostat for controlling a central-heating system is shown in Fig. 26. Increasing the temperature-control setting (anti-clockwise) shifts the moving contact towards the 'on' position. Consequently the bimetallic strip has to bend relatively more in order to open the contacts and break the circuit.

Fig. 26 (*b*) shows a modified version which gives more accurate control of room temperature. A small heating coil inside the thermostat is wired in parallel with the central-heating system; its purpose is to speed up the rise in temperature of the bimetallic strip and thereby anticipate the rise in room temperature.

SUMMARY

All conductors carrying current generate heat: heat and electricity are two forms of energy. Energy is measured in joules. The rate at which energy is converted from one form to another is known as power. A power of 1 watt means an energy conversion of 1 joule per second. Electrical power is the product of current and voltage.

$$P = VI = V^2/R = I^2R \text{ watts}$$

Energy converted (or work done) is the product of power and time.

$$W = Pt = VIt = V^2t/R = I^2Rt \text{ joules}$$

The practical unit of electrical energy is the kilowatt-hour.

$$1 \text{ kWh} = 3{\cdot}6 \times 10^6 \text{ joules}$$

The cost of running an appliance is the product of the current, voltage, time in hours, and the price per kWh.

The speed at which a kilowatt-hour meter turns is proportional to the power drawn from the mains; the total number of revolutions made is proportional to the energy consumed.

Fuses protect appliances by melting and breaking the circuit when an excessive current passes.

Incandescent tungsten-filament lamps have a low (5 per cent) efficiency; fluorescent lamps are relatively more efficient, they depend on the emission visible light from phosphors stimulated by ultraviolet radiation caused by the ionization of mercury vapour.

Thermostats automatically switch off heating appliances at a given temperature; the bimetallic thermostat contains a strip made from two metals which expand by different amounts.

CHAPTER FOUR

ELECTROMAGNETICS

During a lecture demonstration in 1820 the Danish scientist Hans Christian Oersted noticed that a compass needle placed near to a current-carrying wire was deflected from its normal north–south position (Fig. 27). This may not sound a very remarkable discovery, but Oersted realized that it was evidence of a fundamental and far-reaching fact: an electric current is surrounded by a magnetic field.

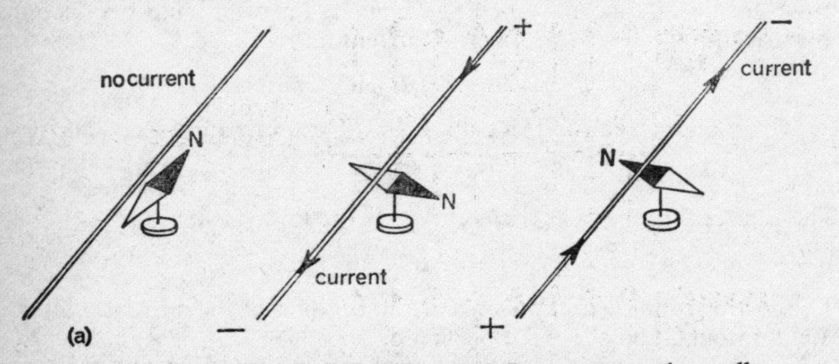

Fig. 27. Oersted's discovery: current deflects a magnetic needle

Before going on to investigate the relationship between electric currents and magnetic fields it will be helpful to digress briefly into a revision of elementary magnetism.

DOMAIN THEORY

Most of the observed facts about magnets can be explained on the assumption that magnetic materials are composed of tiny individual magnets called 'domains'. When a bar magnet is cut in half each half has its own N and S poles. If the halves are themselves broken each resulting fragment is found to possess N and S poles: in fact, the smallest visible fragment into which the magnet can be broken is found to have a pair of poles. Were it possible to continue the breaking process indefinitely, we should end up with a single magnetic domain, complete with N pole and S pole.

A single domain is much too small to be seen (it has been estimated that the head of a pin contains over 6,000 domains), yet it is made up of something like a thousand billion (10^{15}) atoms.

The domains in an ordinary lump of iron are oriented in all directions, as shown in Fig. 28 (*a*), and their magnetic effects cancel out. When the iron is placed near a powerful magnet (i.e. in a magnetic field) the domains begin to rotate—a few at a time—to align themselves with the magnetic field. This is a gradual process that depends on the intensity of the applied field and on the length of time for which it is present.

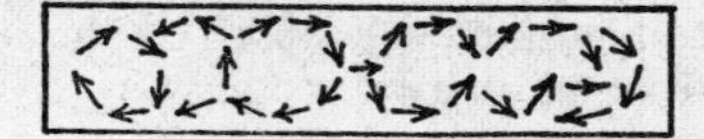

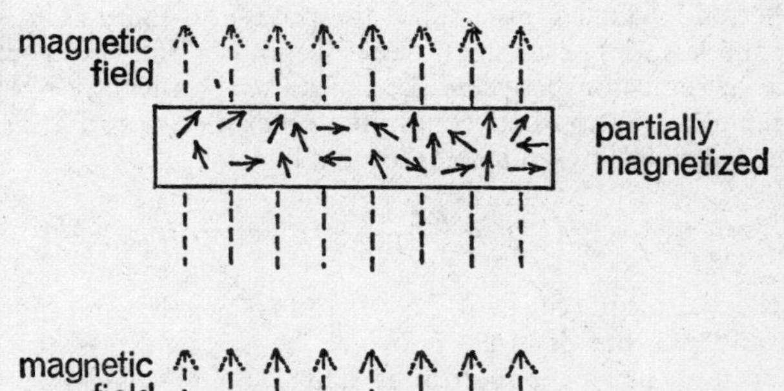

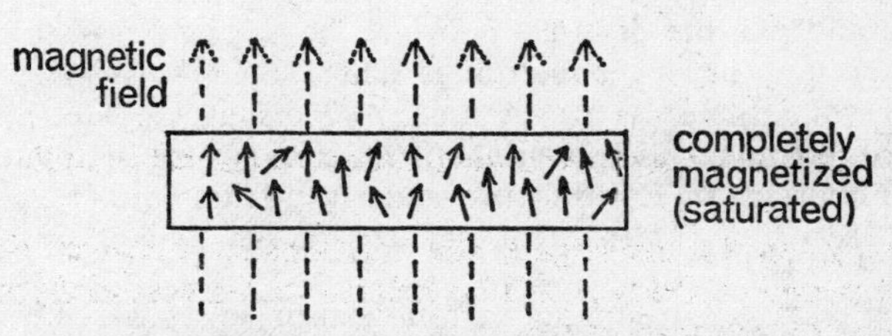

Fig. 28. Alignment of domains when iron is magnetized

Eventually, when all the domains have jumped into line, magnetization is complete and the iron is described as magnetically 'saturated'.

For an explanation of what magnetic domains are, we must look back at our simple picture of the atom (p. 1). The orbital electrons within an atom not only revolve around the nucleus, each one also *spins* about its own axis. In non-magnetic materials the same number of electrons spin clockwise as spin anti-clockwise; consequently, the electron spins in these materials cancel out.

In the atoms of magnetic materials, on the other hand, more electrons spin in one direction than in the other. These uncompensated or unbalanced spins create small magnetic 'twists' called magnetic moments.

Ferromagnetic materials (i.e. iron, steel, and the other metals which are strongly attracted by a magnet) are peculiar in that they contain groups of atoms in which the unbalanced electron spins are all in the same direction—these relatively large groups of atoms are the domains mentioned above. The thousands of parallel spins in a single domain result in intense local 'pockets' of magnetism.

It is an experimental fact that when a magnet is heated to a temperature known as the Curie point its magnetism disappears. This is because the heat absorbed by the magnet makes the atoms vibrate; the alignment of the domains is completely upset by the agitation of the atoms, and they end up oriented in all directions.

The fundamental law of magnetism is that like poles repel and unlike poles attract. 'Poles' are the points near each end of a magnet where the magnetism appears to be concentrated. When a magnet is able to move freely it turns until one end points approximately north and the other end points south; hence the poles of a magnet are distinguished by the letters N (for north-seeking) and S (for south-seeking).

The force of repulsion between like poles (i.e. N and N, or S and S) and the force of attraction between unlike poles (i.e. N and S) are both given by the formula known as Coulomb's law:

$$F = \frac{m_1 \times m_2}{\mu r^2} \qquad (1)$$

where F is the magnitude of the force, m_1 and m_2 are the strengths of the two poles, and r is the distance between them. The 'constant' μ (the Greek letter mu) in this equation is called the *permeability* of the medium in which the poles are situated. As we shall see later, the permeability of ferromagnetic materials (iron, etc.) is very high and μ becomes an important factor in formulas derived from eqn. (1).

MAGNETIC FIELDS

A magnet exerts a force on another magnet or on a piece of iron even when it is some distance away. This mysterious 'action at a distance' is explained by imagining a *field of force* extending throughout the region where the effects of the force can be felt. Although a field is an imaginary concept, it has two real properties that can be measured, namely, its direction and its strength.

The *direction* of a magnetic field is defined by the direction in which an isolated N pole (if it could exist on its own) would tend to move. For example, an isolated N pole situated between the arms of a horseshoe magnet would be attracted to the arm containing the S pole and repelled by the arm containing the magnet's N pole: the direction of the field is thus from the N pole to the S pole.

The *strength* or intensity of a magnetic field can be defined by the force that would be experienced by an isolated N pole of strength 1 unit when placed in the field. (A pole of strength 1 'c.g.s. unit' is one which repels an exactly similar pole, situated 1 cm away in vacuum, with a force of 1 dyne.)

Field strength is then measured in units called oersteds. If a magnetic pole of strength m units experiences a force of F dynes at a point in a magnetic field the strength of the field at that point is

$$H = \frac{F}{m} \text{ oersteds} \tag{2}$$

where H is the symbol used for magnetic field strength or intensity.

According to eqn. (1) on p. 52, the force between two poles is

$$F = \frac{m_1 m_2}{\mu r^2}$$

Hence the force exerted on a pole of strength 1 unit by a magnetic pole of strength m units placed r cm away is

$$F = \frac{m \times 1}{\mu r^2} \text{ dynes}$$

but the force exerted on a pole of strength 1 unit defines the intensity of the magnetic field in which it is situated. Hence a magnetic pole of strength m units produces a field of $\frac{m}{\mu r^2}$ oersteds at any point r cm away. Or, putting it the other way round, at a distance r in a medium of permeability μ the field intensity due to a magnetic pole of strength m units is

$$H = \frac{m}{\mu r^2} \text{ oersteds} \tag{2a}$$

The direction and intensity of a field can be shown by drawing imaginary 'lines of force' to indicate the path that an isolated N pole would take if it were free to move. How close together the lines are drawn depends on the field strength.

A single line of force represents 1 unit of 'magnetic flux'—a rather vague term meaning the magnetism flowing across the field. Flux is symbolized by the Greek letter ϕ (phi) and is measured in units called maxwells. A larger unit of flux called the weber is preferred for practical purposes; one weber $= 10^8$ maxwells.

As mentioned above, the strength of a magnetic field is indicated by the density of its lines of force, i.e. the number of lines per cm^2. This important quantity is known as *flux density*. It is measured in maxwells per cm^2 (also called gauss) or in webers per square metre. One weber per square metre $=$ 10,000 gauss (see Fig. 29).

Using the symbol B for flux density, we can write

$$B = \frac{\phi}{A}$$

or

$$\phi = B \times A$$

where ϕ is the total flux passing through an area A.

No attempt has been made so far to distinguish between flux density B and field intensity H. In a vacuum, B measured in gauss is numerically equal to H measured in oersteds. But flux density is dependent on the

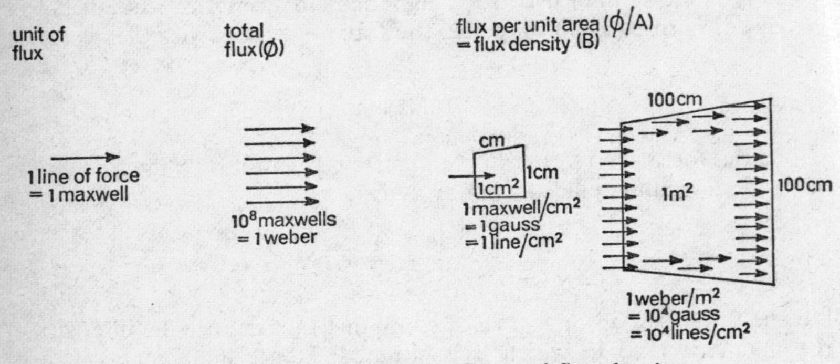

Fig. 29. Units of magnetic flux and flux density

material in which it is produced, whereas field intensity is not. The relationship between B and H is

$$\mu = \frac{B}{H}$$

where μ is the *permeability* of the material, i.e. the ease with which lines of force 'permeate' or pass through it (see p. 59).

ELECTRICAL UNITS

In our digression into the properties of permanent magnets we saw that the 'magnetic domains', which account for most of the observed facts, owe their existence to electron spins. Plainly, a spinning electron represents a movement of electric charge; and as was emphasised in Chapter One, a movement of electric charge is a *current*. Thus we arrive at the conclusion that magnetism is inseparably bound up with electricity: the two are part and parcel of the same phenomenon.

Because every magnetic field is associated basically with current, we need to consider carefully the most convenient units for measuring field strengths, magnetic flux, flux density, and their related quantities.

So far we have considered magnets in isolation and have been content to measure length in centimetres: in fact we have employed the C.G.S.

system of units. In the C.G.S. system the unit of force is the dyne, the unit of magnetic field strength is the oersted, the unit of magnetic flux is the maxwell, and the unit of flux density is the gauss.

However, the C.G.S. system has certain disadvantages where electrical quantities are concerned. The practical units of current and e.m.f. are—as we know from previous chapters—the ampere and the volt. But these practical units do not fit into the C.G.S. system unless conversion factors are taken into account.

On the other hand, when lengths are measured in metres, mass in kilograms, and time in seconds we have a system whose unit of current really is the ampere and whose unit of e.m.f. really is the volt. This is known as the M.K.S. system, and is used (in a slightly modified form called the rationalized M.K.S. system) throughout electrical engineering. It is accepted internationally as the Système International or SI unit system.

In the M.K.S. system the unit of force is the newton (N), the unit of magnetic flux is the weber (Wb), the unit of flux density is the weber per square metre (Wb/m^2) or tesla, and the unit of magnetic field strength is the ampere-turn per metre (At/m). These units are listed with their C.G.S. equivalents in Table 5 on p. 91.

FIELD ROUND A CONDUCTOR

An electric current is, as Oersted observed, surrounded by a magnetic field. The lines of force around a straight current-carrying wire (Fig. 30) consist of concentric circles which are spaced farther apart at increasing distances from the wire. Fig. 30 shows that when the (conventional)

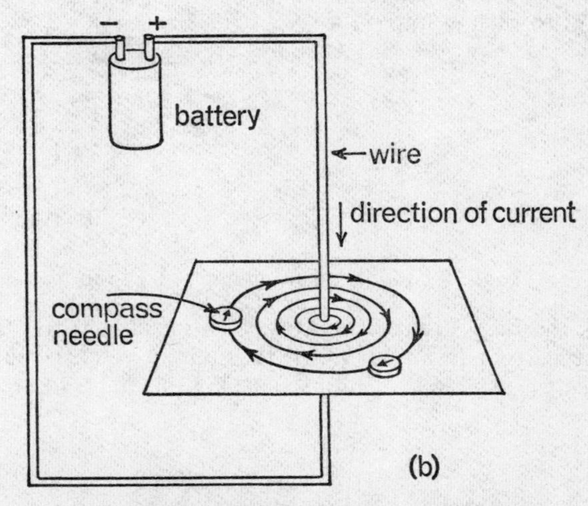

Fig. 30. Magnetic field around a straight wire

current is flowing away from the observer the direction of the magnetic field is clockwise; conversely, when the current flows towards the observer the direction of the field is anti-clockwise.

The intensity of the field around a long straight wire is related to the current in the wire by the formula

$$H = \frac{I}{2\pi r} \tag{3}$$

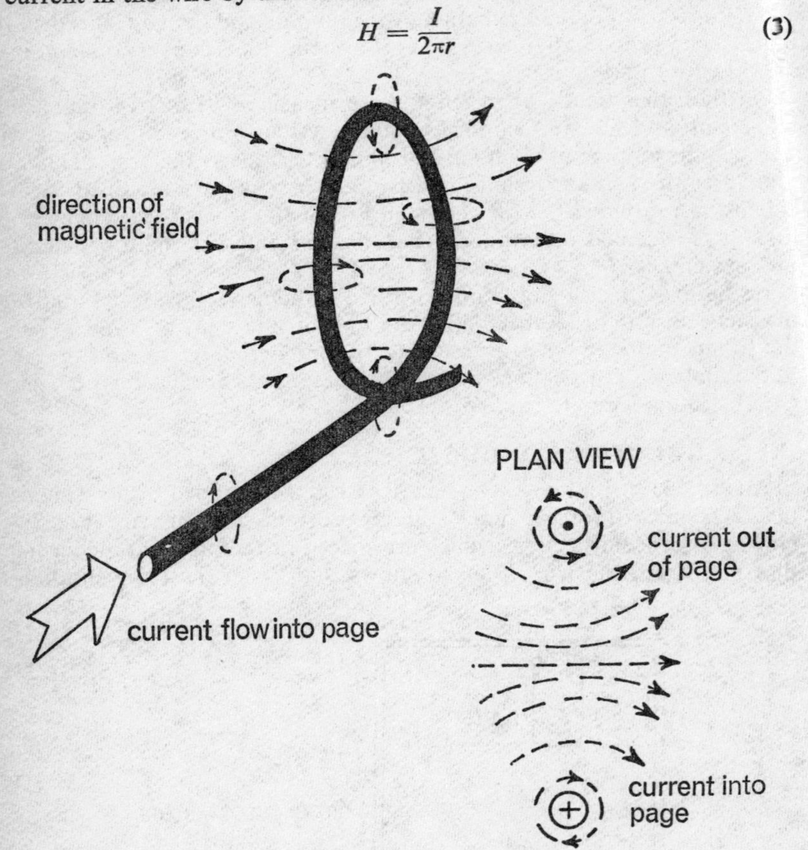

Fig. 31. Magnetic field in a circular loop

where H is the field strength in At/m at a point r m from the wire, and I is the current in amps. When the same wire carrying the same current is bent into a circle of radius r m (Fig. 31) the field at the centre of the loop is

$$H = \frac{I}{2r} \tag{4}$$

Since π stands for $\frac{22}{7}$ or 3·14159, the simple expedient of coiling the wire into a loop increases the magnetic-field strength by rather more than a factor of 3.

As might be expected, a double loop gives twice the field strength of a single loop, and a coil of n turns has at its centre a field strength n times that of a single loop, i.e.

$$H = \frac{nI}{2r} \text{ ampere-turns per metre} \tag{5}$$

Eqn. (5) is the formula for the field at the centre of a short coil, i.e. one whose length is small compared with its diameter. It is not valid for

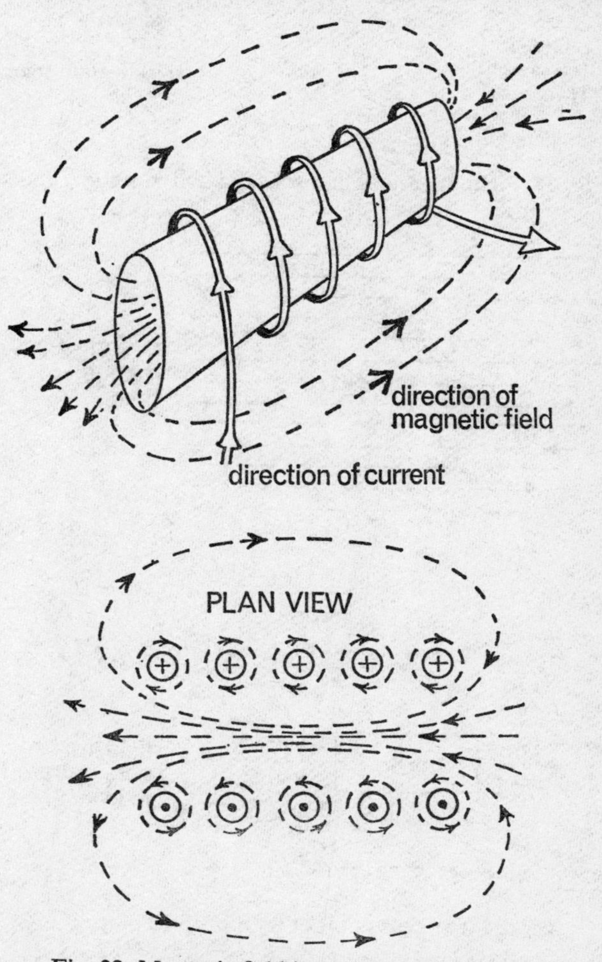

Fig. 32. Magnetic field in and around a long coil

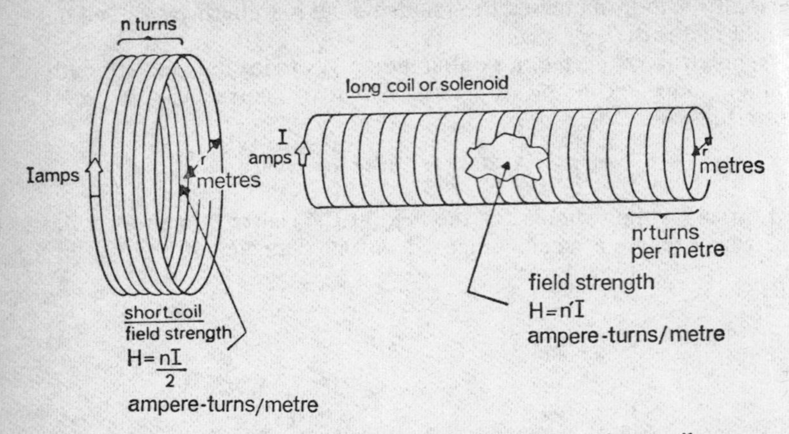

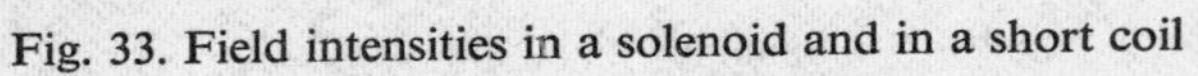
Fig. 33. Field intensities in a solenoid and in a short coil

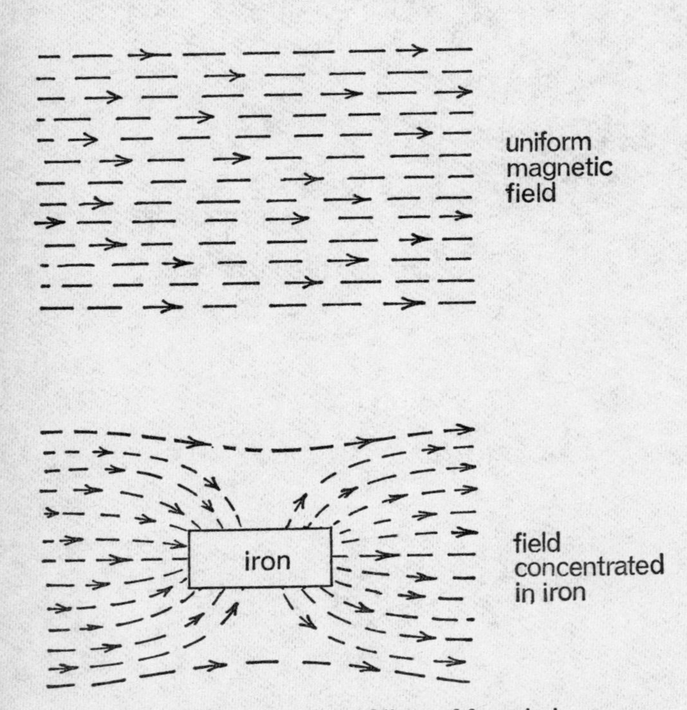

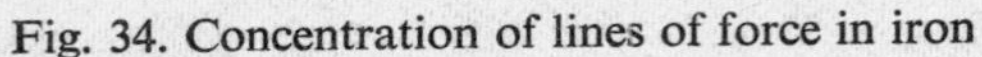
Fig. 34. Concentration of lines of force in iron

a long coil or solenoid, i.e. one whose radius is small compared with its length, since the lines of force at the centre of such a coil are parallel and evenly spaced, as shown in Fig. 32. The intensity of this uniform field is

$$H = \frac{nI}{l} \text{ ampere-turns per metre} \tag{6}$$

where n is the total number of turns, I is the current in amps, and l is the length of the coil in metres.

The ratio $\frac{n}{l}$ is the number of turns per metre, and is written as n'. Eqn. (6) may then be modified to

$$H = n'I \text{ ampere-turns per metre} \tag{7}$$

Eqn. (7) also gives the intensity of the field in a toroid, i.e. a coil that is bent into a ring rather like a doughnut. A toroid has no ends and there is no leakage of flux: the magnetic field is entirely confined within the loops of wire.

Compared with a permanent magnet, a solenoid carrying a current produces remarkably little magnetic flux. But the output can be increased enormously, without altering the current, by inserting an iron core into the coil.

Iron has a permeability several thousand times that of air or empty space, and since permeability (the relative ease with which lines of force pervade a material) is defined by the ratio of flux density B to field intensity H (i.e. $\mu = B/H$), it should be apparent that the flux density increases several thousand times when a core of soft wrought iron is pushed into a solenoid.

Fig. 34 shows how a uniform magnetic field is distorted by the introduction of a piece of iron. Iron has a high permeability, and the lines of force tend to concentrate in it. Temporarily, the iron is a magnet and produces additional lines of force.

ELECTROMAGNETS

A coil wound on an iron core is called an electromagnet, and the field around an electromagnet closely resembles that around a bar magnet, as shown in Fig. 35. But there is one important difference between an electromagnet and a permanent magnet: when the current is cut off the magnetic field of an electromagnet at once disappears almost completely. This ability to switch the field on and off as required is extremely useful in practice.

One device which utilizes the switching facility of an electromagnet is the relay (Fig. 36) used in telephone exchanges, counting mechanisms, and all kinds of control and automation apparatus. Basically, the relay is an electromagnet mounted close to a lever (the armature) made of some ferromagnetic material. When current flows in the windings of the electromagnet the armature is attracted towards the core,

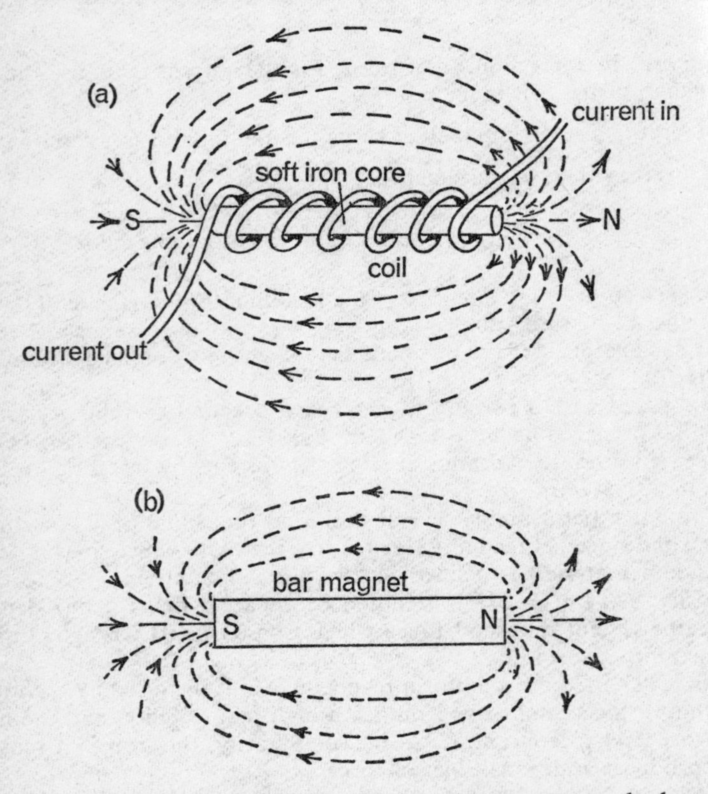

Fig. 35. Comparison of fields around an electromagnet and a bar magnet

swings on its pivot, and in doing so closes a pair of spring-loaded contacts.

As soon as the current stops, the core ceases to attract; the armature swings back to its original position, and the contacts open. The importance of this system lies in the fact that the current in the electromagnet is completely separate from the current carried by the contacts. Thus a weak current in the electromagnet can switch a heavy current through the contacts. The contacts shown in Fig. 36 are 'normally open'; they could easily be arranged to be 'normally closed', so that a weak current in the electromagnet would open the contacts and switch off a heavy current.

A more familiar application of the electromagnet is the electric bell. The bell illustrated in Fig. 37 has two coils connected in series to form a horseshoe type of magnet. A strip of iron held just in front of the ends cf the horseshoe by a leaf spring acts as the armature. In contrast to the relay, the armature of a bell does form part of the electrical circuit.

Pressing the bell-push completes the circuit; current flows from the

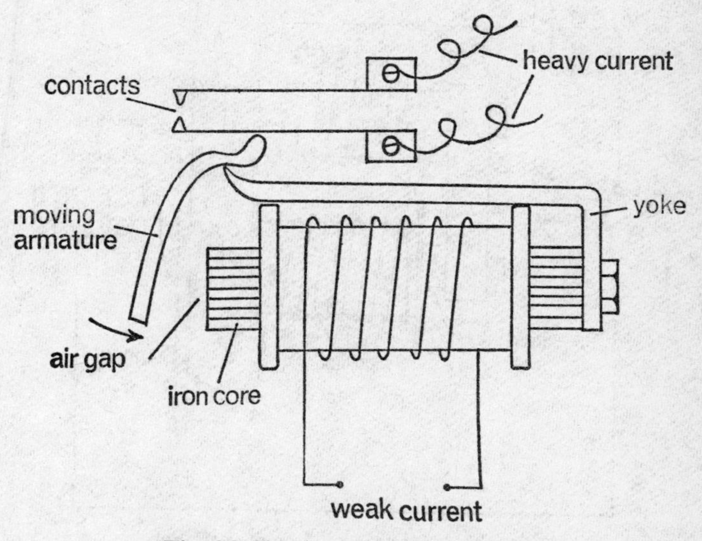

Fig. 36. Simple electromagnetic relay

battery, through both windings of the electromagnet, and into the armature. From the armature it passes through a pair of contacts, one of which is attached to the armature by a spring-steel strip. The other is mounted on an adjusting screw so that the gap between the contacts may be altered in order to vary the tone of the bell. Current flows from the adjustable contact via the bell-push back to the battery.

When the armature is attracted towards the electromagnet it carries one of the contacts with it, and thus breaks the circuit; it also causes the hammer to strike the bell. The magnet no longer attracts the armature, which springs back to its original position, and the contacts close to complete the circuit once more. This action is rapidly repeated, making the hammer hit the bell many times every second.

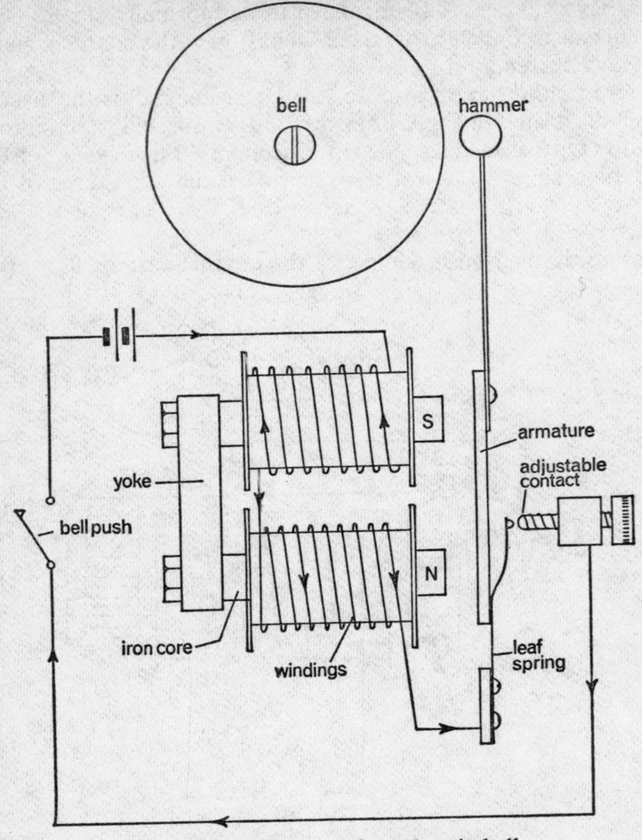

Fig. 37. Mechanism of an electric bell

MAGNETIC CIRCUITS

So far no mention has been made of the 'yoke' of the electromagnets illustrated in Figs. 36 and 37. Although it might appear unnecessary at first sight, in fact it plays an important role in raising the efficiency of an electromagnet. It provides an 'easy' path for the magnetic flux, without which the force of attraction between the magnet and the armature would be much reduced.

Just as an electric current flows more readily in a good conductor, such as copper, than in, say, a glass rod, so magnetic flux passes more readily through iron ($\mu_r = 5{,}000$) than through, say, air ($\mu_r = 1$)*. The

* One of the penalties we have to pay for abandoning the C.G.S. system is that the permeability of free space (i.e. a vacuum) is no longer $\mu = 1$. In the rationalized M.K.S. system, the permeability of free space is $\mu_o = 4\pi \times 10^{-7}$. The permeability of any material compared with that of free space is called the *relative permeability* and is symbolized by μ_r. Being a ratio, μ_r has no units to be taken into account. The *absolute permeability* of any material is the product $\mu_r\mu_o$.

analogy between magnetic flux and electric current can be carried further, and we can speak of magnetic 'circuits' and even impose a form of Ohm's law on them. (However, it should be remembered that in a magnetic circuit there is no actual movement corresponding to the flow of electrons: flux is 'carried' by lines of force which are, of course, imaginary.)

In the same way that magnetic flux (ϕ) corresponds to an electric current (I), so magnetomotive force (F) is the counterpart of electromotive force (E). The magnetic counterpart of electrical resistance (R) is called reluctance (R), and just as Ohm's law applies in the form $I = \frac{E}{R}$ to electrical circuits, we can apply the corresponding equation

$$\phi = \frac{F}{R} \qquad (8)$$

to a magnetic circuit. If the flux ϕ is in webers the magnetomotive force F is in ampere-turns. The unit of reluctance has no special name.

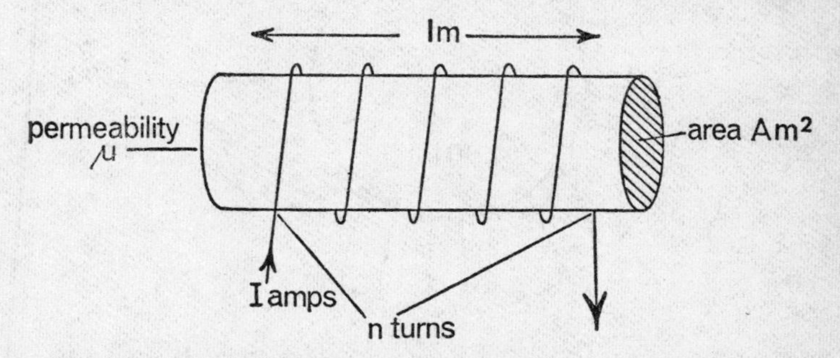

Fig. 38. Formula for the m.m.f. produced by an electromagnet

Furthermore, just as electrical resistance is related to the shape and material of a conductor in accordance with the formula $R = \rho l/A$, so magnetic reluctance depends on the shape and material of the magnetic circuit. The formula for this relationship is

$$R = \left(\frac{1}{\mu}\right)l/A = \frac{l}{\mu A} \qquad (9)$$

when μ ($= \mu_r\mu_0$) is the permeability of the material, l is the length of the circuit, and A is the area of cross-section.

The magnetomotive force (or m.m.f.) produced by any electromagnet can be found from the definition of the ampere-turn as the unit of magnetomotive force. One ampere-turn (1 At) is the m.m.f. produced by a current of 1 A flowing in a coil of one turn: hence the m.m.f. produced by a current I amps flowing in a coil of n turns is

$$F = nI \text{ ampere-turns} \qquad (10)$$

Suppose the electromagnet is l metres long, has a cross-sectional area of A square metres, has n turns, carries a current of I amps and is wound on a core of permeability μ, as shown in Fig. 38.

Turning round eqn. (8), we can write

$$\begin{aligned} \text{m.m.f.} &= \text{Flux} \times \text{Reluctance} \\ &= B \times A \times \text{Reluctance} \quad (\text{since } B = \phi/A) \\ &= \mu HA \times \text{Reluctance} \quad (\text{since } \mu = B/H) \\ &= \mu HA \times \frac{l}{\mu A} \quad (\text{from eqn. (9)}) \\ &= H \times l \end{aligned}$$

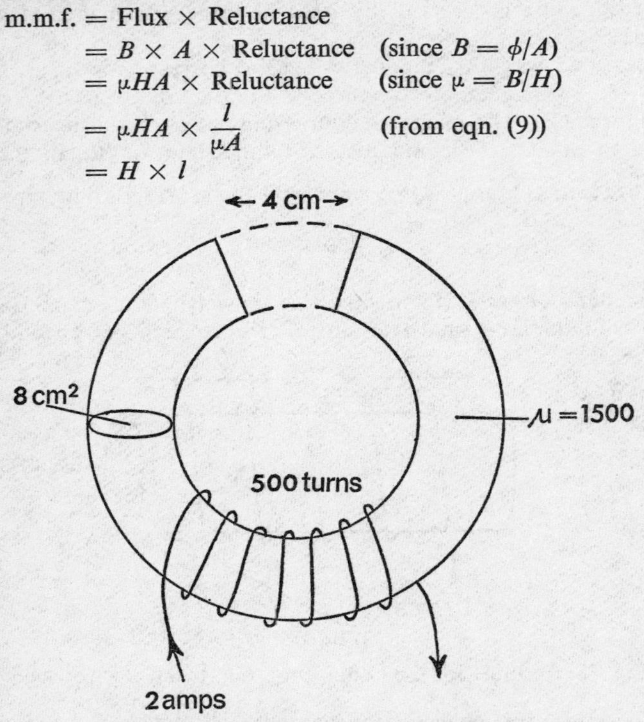

Fig. 39. Magnetic circuit with air gap

Hence the m.m.f. is the product of field strength H and length l of the coil, or

$$\text{Field strength} = \frac{\text{m.m.f.}}{\text{Coil length}}$$

$$H = \frac{F}{l} \tag{11}$$

If F is in ampere-turns and l is in metres, H will be in ampere-turns per metre (At/m). Thus eqn. (11) defines the M.K.S. unit of magnetic field strength, also called the magnetizing force.

EXAMPLE: The core of the electromagnet shown in Fig. 39 is a ring whose average circumference is 40 cm and whose area of cross-section is 8 cm²; the relative permeability of the core is $\mu_r = 1{,}500$. The coil has 500 turns and carries a current of 2 A. What is the total flux in the core?

From eqn. (10) the m.m.f. is

$$F = nI$$
$$= 500 \times 2$$
$$= 1{,}000 \text{ ampere-turns}$$

From eqn. (9) the reluctance of the core is

$$R = \frac{l}{\mu A}$$
$$l = 40 \text{ cm} = 0{\cdot}4 \text{ m}$$
$$A = 8 \text{ cm}^2 = 8 \times 10^{-4} \text{ m}^2$$
$$\mu = \mu_r \mu_o = 1{,}500 \times 4\pi \times 10^{-7}$$
$$R = \frac{0{\cdot}4}{1{,}500 \times 4\pi \times 10^{-7} \times 8 \times 10^{-4}}$$
$$= \frac{10^7}{12\pi}$$

From eqn. (8) the total flux is

$$\phi = \frac{F}{R}$$
$$= \frac{1{,}000 \times 12\pi}{10^7}$$
$$= 0{\cdot}00377 \text{ webers} \quad \textit{Answer}$$

EXAMPLE: If a piece is removed from the ring in the previous example, leaving an air gap 4 cm long, what is the total flux in the core?

The m.m.f. is still $F = 1{,}000$ ampere-turns.

The reluctance of the core is made up of two parts: R_{iron} for the 36 cm length of iron, and R_{air} for the 4 cm air gap.

$$R_{iron} = \frac{l}{\mu A}$$
$$l = 36 \text{ cm} = 0{\cdot}36 \text{ m}$$
$$A = 8 \text{ cm}^2 = 8 \times 10^{-4} \text{ m}^2$$
$$\mu = \mu_r \mu_o = 1{,}500 \times 4\pi \times 10^{-7}$$
$$R_{iron} = \frac{0{\cdot}36}{1{,}500 \times 4\pi \times 10^{-7} \times 8 \times 10^{-4}}$$
$$= 238{,}700$$
$$R_{air} = \frac{l}{\mu A}$$
$$l = 4 \text{ cm} = 0{\cdot}04 \text{ m}$$
$$A = 8 \times 10^{-4} \text{ m}^2$$
$$\mu_r = 1 \text{ for air}$$
$$\mu = \mu_r \mu_o = 4\pi \times 10^{-7}$$
$$R_{air} = \frac{0{\cdot}04}{4\pi \times 10^{-7} \times 8 \times 10^{-4}}$$
$$= 39{,}800{,}000$$

The iron core and the air gap are in series, and series reluctances may be added together just like series resistances. Hence the total reluctance is

$$\begin{aligned} R &= R_{\text{iron}} + R_{\text{air}} \\ &= 238{,}700 + 39{,}800{,}000 \\ &= 4 \times 10^7 \end{aligned}$$

The total flux is

$$\begin{aligned} \phi &= \frac{F}{R} \\ &= \frac{1{,}000}{4 \times 10^7} \\ &= 0{\cdot}000025 \text{ webers} \quad \textit{Answer} \end{aligned}$$

Thus the air gap reduces the total flux by a factor of 150, and we can now see why the yoke is a necessary part of an electromagnet and why air gaps in magnetic circuits should be kept as small as possible.

HYSTERESIS

The term 'permeability' has appeared several times in this chapter and has been defined as the ratio of flux density B to field intensity H. The

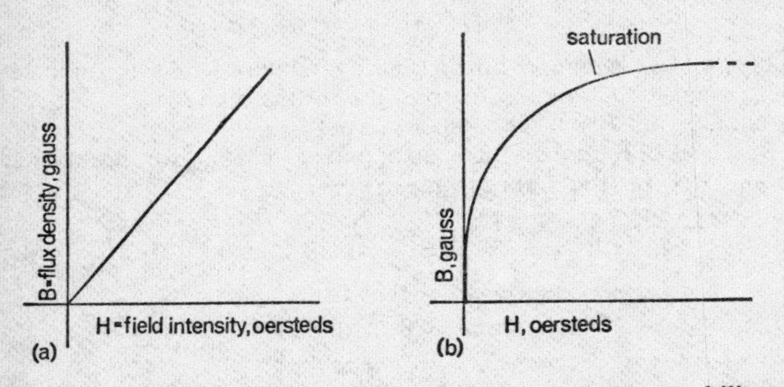

Fig. 40. Magnetization curves: (*a*) for constant permeability; (*b*) experimental result for iron

formula $\mu = B/H$ or $B = \mu H$ might suggest that permeability is a constant for any given substance. Unfortunately this is not true for the important ferromagnetic materials. If μ were constant a graph of flux density plotted against field intensity would be a straight line, as in Fig. 40 (*a*). Experimental results for a specimen of iron give a curved B/H graph, as in Fig. 40 (*b*). The reason for this unexpected behaviour lies with the magnetic 'domains' discussed on p. 50.

When a specimen of iron or other ferromagnetic material is placed in a magnetic field the domains jump into line. If the magnetizing field is strong enough the gradual process of alignment goes on until all the domains are parallel and the specimen is 'saturated', as shown in Fig. 28.

The magnetization of the specimen can conveniently be represented by the flux density B produced in it. If the flux density is measured for various values of the magnetizing-field intensity H, the results give a magnetization curve similar to that shown in Fig. 40 (*b*). It will be seen that the flux density increases as the field is increased, rapidly at first (where the curve shows a steep slope) but much more slowly near saturation. Finally, the curve levels off* and the flux density remains practically fixed, no matter how much the magnetizing field is increased.

The magnetization curve is obviously not a straight line (except when B and H are very small) and the permeability ($\mu = B/H$) is not constant. We can determine the value of μ for any given intensity H of the magnetizing field by measuring the ratio B/H at the point corresponding to the given value of H.

Increasing the magnetizing field, as described above, until saturation occurs is only half the story. The sequel begins when the field is reduced again in order to demagnetize the specimen: it is found that when $H = 0$ the flux density is by no means zero.

The full story is shown in Fig. 41. Starting with no field ($H = 0$) and a completely unmagnetized specimen ($B = 0$), the field is increased until saturation occurs at point *a*. Curve *o–a* is essentially the same as Fig. 40 (*b*). The field is now reduced gradually to zero, and it is found that the flux density does not retrace the original curve *o–a* but follows a higher curve *a–b*. At point *b* the field is zero; there is, however, a considerable flux density (*o–b* in the diagram) remaining in the specimen. This residual flux is known as *remanence.*

In order to demagnetize the specimen, i.e. to bring B back to zero, the magnetizing field has to be reversed. The reversed field is gradually increased until $B = 0$, which occurs at point *c*. The value (*o–c*) of the reversed field required to demagnetize the specimen completely is known as *coercive force.*

Continuing the gradual increase of the reversed field causes the specimen to be magnetized in the opposite direction (i.e. with poles reversed) until saturation is reached at point *d*. When the reversed field is reduced again to zero at point *e* the residual flux density is *o–e*. To demagnetize the specimen once more, it is necessary to reverse the field a second time (so that it is in the original direction) and increase it until, at point *f*, the flux density is zero.

With further increases in field intensity, the flux density follows the curve *f–a* until saturation occurs. If the field is carried through another cycle (reduce–reverse–increase–reduce–reverse–increase, etc.), the flux density will once again follow the curve *abcdef*, etc.

* Actually the curve continues to rise very slightly, the relatively small slope being identical with that if the material were air.

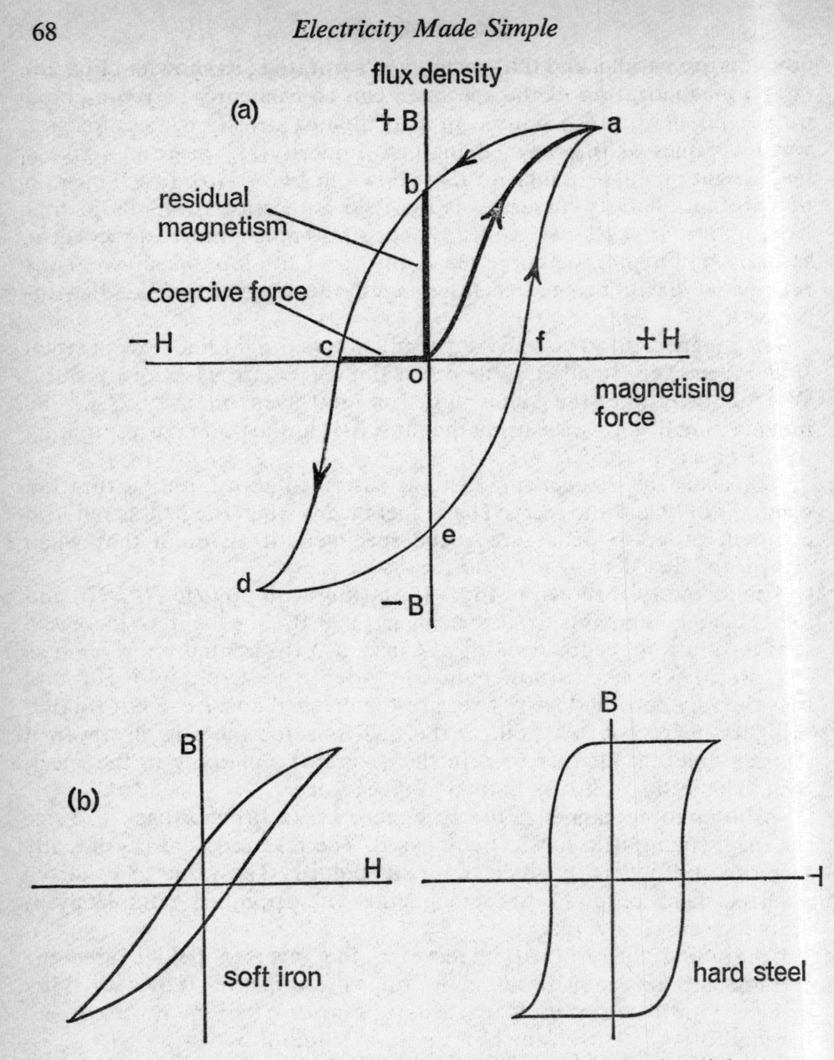

Fig. 41. Hysteresis loops for typical ferromagnetic materials

In the above description the flux density in the specimen is seen to be lagging behind the magnetizing field. This lag of B behind H is called *magnetic hysteresis*, and the closed curve *abcdef* (Fig. 41 (*a*)) is called a *hysteresis loop*. The area enclosed by the hysteresis loop is a measure of the energy wasted (converted to heat) in magnetizing and demagnetizing a material. The wasted energy is known as *hysteresis loss*.

Fig. 41 (*b*) shows a narrow hysteresis loop for soft wrought iron and a broad hysteresis loop for 'hard' steel. Magnetically hard materials (steel and various alloys) are characterized by a loop of large area, signifying a large remanence (residual magnetism) and the need for a large coercive force to demagnetize the material. 'Hard' materials are suitable for permanent magnets, which must store large amounts of magnetic energy and resist the effects of demagnetizing fields.

The loop for soft iron, on the other hand, has a small area, signifying small residual magnetism and small coercive force. 'Soft' materials waste little energy in 'hysteresis losses' (see p. 127), and are therefore suitable for the cores of alternating-current motors, generators, transformers, etc.

SUMMARY

Ferromagnetic materials are those such as iron and steel which are strongly attracted to a magnet. They contain groups of atoms, called domains, which jump into line when the material is placed in a magnetic field. The domains act as minute individual magnets.

Magnetism is always associated with current. The region in which magnetic influence or *flux* can be detected is known as a *field.*

Magnetic fields can be considered as circuits analogous to electrical circuits: in this analogy the formula $\phi = \frac{F}{R}$ is the counterpart of Ohm's law $I = \frac{E}{R}$, where the *magnetomotive force* F corresponds to electromotive force E, flux ϕ corresponds to current I, and reluctance R corresponds to resistance. Magnetic flux is measured in webers (Wb) or in maxwells.

In the rationalized M.K.S. system the unit of magnetomotive force is the ampere-turn (At): a current of I amps flowing in a coil of n turns produces an m.m.f. $F = nI$ ampere-turns. If the length of the coil is l metres the m.m.f. is $F = Hl$ ampere-turns, where H is the magnetic field strength of magnetizing force. Hence

$$H = \frac{F}{l} = \frac{nI}{l}$$

and is measured in ampere-turns per metre (At/m).

Any material placed in a magnetic field has magnetic flux induced in it. The *density* of this flux, i.e. flux/area $= \phi/A$, depends on the magnetic field strength and on the magnetic *permeability* of the material. These quantities are related by the formula

$$B = \mu H$$

where B is the flux density in webers per square metre (Wb/m^2), H is the field strength in At/m and μ is the permeability of the material.

$\mu = 4\pi \times 10^{-7}$ for free space (vacuum) and has practically the same value for air.

Permeability of ferromagnetic materials varies with the intensity of the magnetizing field. Changes in flux density lag behind changes in the intensity of the field producing them: this lag is called magnetic hysteresis.

CHAPTER FIVE

ELECTROCHEMISTRY

It is well known that chemical action can produce an e.m.f. and hence an electric current—for this is precisely what goes on in a torch battery. The converse is equally true: an electric current can cause a chemical action. The latter process, though not so familiar to many people, is of tremendous importance in industry. It plays a vital part in the extraction from their ores of aluminium, sodium, calcium, magnesium, etc., and it is the basis of electroplating of metals with nickel, chromium, etc.

Pure liquids, except liquid metals such as mercury, are as a rule non-conductors of electricity (mineral oil, for example, is a very good insulator and is used as such in transformers (p. 129)). But liquids in which a salt (e.g. common salt, sodium chloride) has been dissolved will conduct electricity.

IONS

When a current passes through a wire no observable change occurs: the metal wire is just the same after carrying a current as it was before. When a salt solution carries a current, however, a change does occur, usually leading to chemical decomposition; this is known as *electrolysis.*

The big difference between conduction in a metal and conduction in a solution is that in the former current is carried by electrons, while in the solution (or, to give it its proper name, electrolyte) current is carried by *ions*. Ions are atoms that have either gained or lost one or more of their outermost electrons. An atom that gains an electron gains a negative charge and is called an *anion.* An atom that loses an electron gains a positive charge and is called a *cation.* The conductors dipping into the electrolyte so that it can be connected to a battery (Fig. 42) are called *electrodes*; the positive connexion is the *anode* and the negative connexion is the *cathode.*

Anions, being negatively charged, are attracted to the positive anode; cations, being positively charged, are attracted to the negative cathode.

The ions are present in the solution whether the current is present or not; in fact, even in the solid crystalline state a compound such as sodium chloride is built up from positive and negative ions rather than neutral atoms.

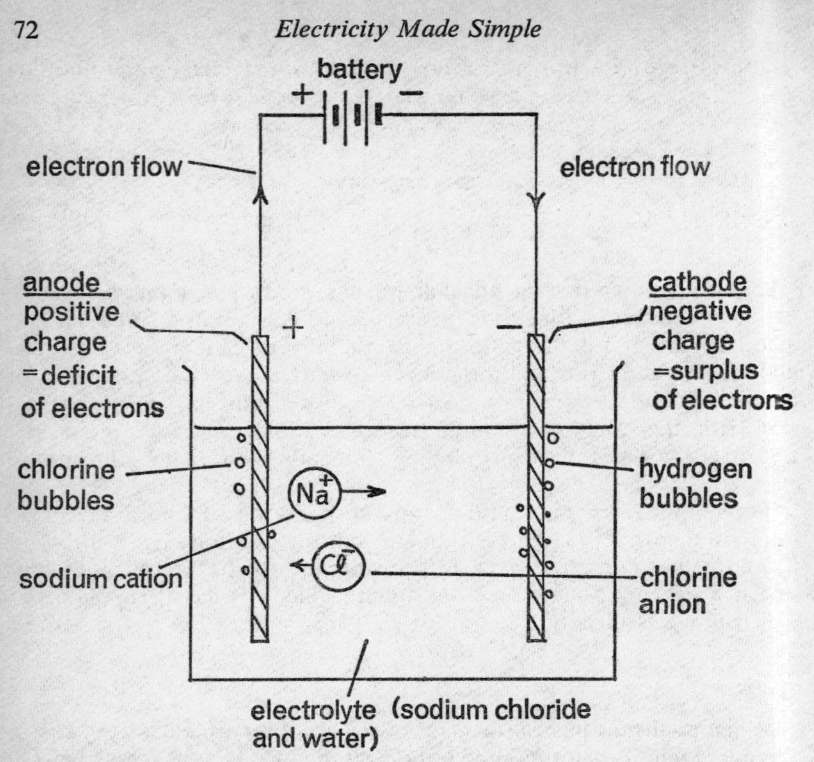

Fig. 42. Electrolysis of a solution of sodium chloride (common salt) in water

ELECTROLYSIS OF SODIUM CHLORIDE

In the experiment shown in Fig. 42 it is apparent that some kind of chemical reaction is taking place: gas bubbles are seen streaming to the surface of the electrolyte at both the anode and the cathode. Initially, the salt solution has no effect on litmus paper (it is neutral), but after the experiment has gone on for some time red litmus is turned blue in the region of the cathode.

Assuming that the electrodes are chemically inert (they may be platinum wires or carbon rods), what is the chemistry of the situation described above? The electrolyte contains water, positive sodium ions, and negative chloride ions. The sodium cations are naturally attracted to the negative cathode, and an equal number of chloride anions are attracted to the anode.

At the cathode, which is connected to the negative terminal of the battery, there is a negative charge, i.e. a *surplus* of electrons; at the anode, connected to the positive battery terminal, there is a *deficit* of electrons.

Now a chloride ion is an atom of chlorine that has gained one extra electron: its symbol is Cl^-. At the anode the chlorine ion gives up its surplus electron in an attempt to cancel the electron deficit on the anode, and thereby becomes a normal atom of chlorine. In practice, chlorine atoms occur in pairs (making a molecule of chlorine, symbol Cl_2), and so two chloride ions surrender their surplus electrons to the cathode:

$$\underset{\text{Ions}}{2Cl^-} \longrightarrow \underset{\text{Molecule}}{Cl_2} + \underset{\text{Electrons}}{2e^-}$$

where e^- is the symbol for an electron. This molecule of chlorine joins up with innumerable similar molecules to form a bubble of chlorine gas which rises to the surface and escapes.

The reaction at the cathode is not quite so simple. In theory, the sodium ion ought to receive one of the surplus of electrons on the cathode and thereby become a normal atom of sodium:

$$\underset{\text{Ion}}{Na^+} + \underset{\text{Electron}}{e^-} \longrightarrow \underset{\text{Atom}}{Na}$$

In fact, sodium is not formed; sodium is a very reactive metal and reacts vigorously with water to produce sodium hydroxide (caustic soda) and hydrogen:

$$\underset{\text{Sodium}}{2Na} + \underset{\text{Water}}{H_2O} \longrightarrow \underset{\substack{\text{Sodium}\\\text{hydroxide}}}{2NaOH} + \underset{\text{Hydrogen}}{H_2}$$

What actually happens is that sodium ions react with water, stealing the $(OH)^-$ they need to form sodium hydroxide, leaving hydrogen H^+ ions which receive electrons from the cathode to become hydrogen atoms. The net result of the reaction is the same as that indicated above: caustic soda is formed in the electrolyte, and hydrogen gas is produced on the cathode. The hydrogen gas can be seen as bubbles rising to the surface, and the presence of caustic soda can be detected by its action in turning red litmus paper blue.

Incidentally, if the electrolysis were done not with a solution of common salt in water, but with molten sodium chloride, sodium metal would be liberated at the cathode, since there would be no water for it to react with.

ELECTROLYSIS OF WATER

Water is the most common of all liquids: it is also the best of all solvents, and for this reason is hardly ever pure. Pure water is a bad conductor of electricity, but tapwater contains dissolved salts which make it conduct (hence it is potentially dangerous to handle electrical apparatus in the presence of water). The electrolysis of water is generally demonstrated after adding a small quantity of dilute sulphuric acid to ensure that enough ions are present to carry an effective current (Fig. 43).

Sulphuric acid (H_2SO_4) molecules split up in water into a sulphate anion (SO_4^{--}) and two hydrogen cations (H^+):

$$\underset{\text{Acid molecule}}{H_2SO_4} \longrightarrow \underset{\text{Hydrogen ions}}{2H^+} + \underset{\text{Sulphate ion}}{SO_4^{--}}$$

Unlike the hydrogen ion or the chloride ion described above, the sulphate ion is *doubly* charged: it has a surplus of *two* electrons. Being negative, it is attracted to the anode.

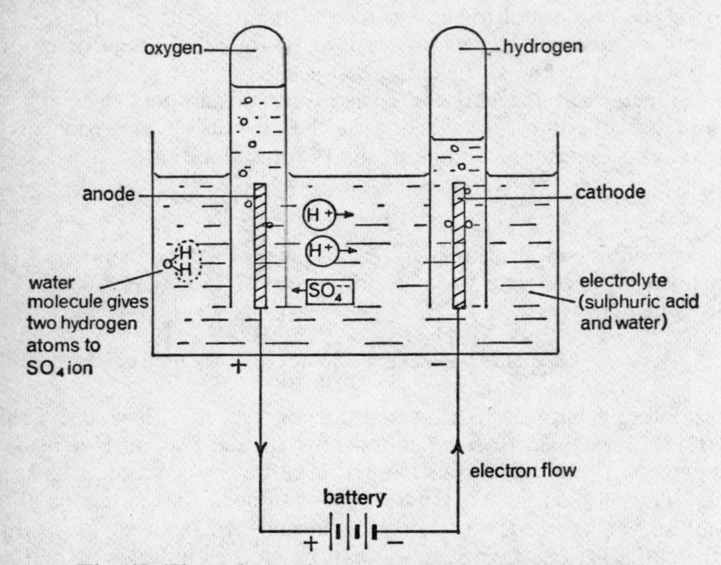

Fig. 43. Electrolysis of water containing sulphuric acid

At the anode each sulphate ion steals a pair of hydrogen ions from a water molecule to form a new sulphuric acid molecule. The debris of the broken water molecule, which, for simplicity, we shall call an oxygen ion (O^{--}), is able and willing to give up its two surplus electrons to the positive charge (i.e. electron deficit) on the anode and thus become a neutral atom of oxygen. Each atom formed in this way links up with a second atom to become an oxygen molecule, and ultimately a bubble of oxygen gas. Summarizing these reactions:

$$\underset{\text{Sulphate ion}}{SO_4^{--}} + \underset{\text{Water molecule}}{H_2O} \longrightarrow \underset{\text{Acid molecule}}{H_2SO_4} + \underset{\text{Atom of oxygen}}{\tfrac{1}{2}O_2} + \underset{\text{Electrons}}{2e^-}$$

Hence the overall result is the evolution of oxygen at the anode. There is no net change in the number of sulphuric acid molecules, since for the one originally broken up another has been formed.

At the cathode there is a surplus of electrons, and hence a negative charge, which attracts the positively charged hydrogen ions formed by the break-up of sulphuric acid molecules.

Each hydrogen ion combines with an electron from the cathode to form an atom, and each pair of atoms combine to form a hydrogen molecule:

$$2H^+ + 2e^- \longrightarrow H_2$$

The electrons are replaced on the cathode by 'new' ones drawn from the negative terminal of the battery; meanwhile, enough hydrogen molecules have been formed to produce a bubble of hydrogen gas.

If the gases given off at the electrodes are collected it will be seen that twice as much *h*ydrogen is formed (at the cat*h*ode) as *o*xygen at the an*o*de. This is what we should expect from the theory discussed above: the equations show that for every hydrogen molecule only half a molecule of oxygen is produced.

The sulphuric acid is not used up—although it plays an important part in the reaction—and can be regarded as a catalyst.

ELECTROPLATING

In the two examples of electrolysis considered so far the electrodes are inert and do not enter into the reaction. Electrodes do enter into the reaction, and usefully so, in the process of electroplating. As an example, let us consider an easy experiment in which an iron key is plated with copper (Fig. 44).

The key to be plated is connected to the negative terminal of the battery so it becomes the cathode. A sheet of copper is connected to the positive terminal so that it becomes the anode. The electrolyte is a solution of copper sulphate (other copper salts could be used) made by dissolving in water as many copper sulphate crystals as possible.

The battery shown in Fig. 44 has an e.m.f. of about 4·5 V; this is not critical, it has been chosen so that the plating process is reasonably rapid. After the circuit has been connected up for only a few minutes the key is found to be completely covered with copper. The longer the current lasts, the thicker will be the coating—but a more even deposit is obtained using a small current for a long time rather than a heavier current for a shorter period.

Incidentally, the plating on the key is *pure* copper, even if the anode is made of very impure copper: hence this is a method of purifying metals.

The reactions going on in the electroplating experiment are as follows. As soon as the copper sulphate crystals are dissolved in water they split up into positive copper ions (Cu^{++}) and negative sulphate ions

$$CuSO_4 \longrightarrow Cu^{++} + SO_4^{--}$$

each of which is doubly charged.

The positive copper cations are attracted by the negative charge, i.e. surplus of electrons, on the iron key connected to the negative battery terminal. When they reach the key each cation combines with two electrons, and an atom of copper is deposited:

$$\underset{\text{Ion}}{Cu^{++}} + \underset{\text{Electrons}}{2e^-} \longrightarrow \underset{\text{Atom}}{Cu}$$

The negative sulphate anions are attracted by the positive charge (deficit of electrons) on the copper strip connected to the positive battery

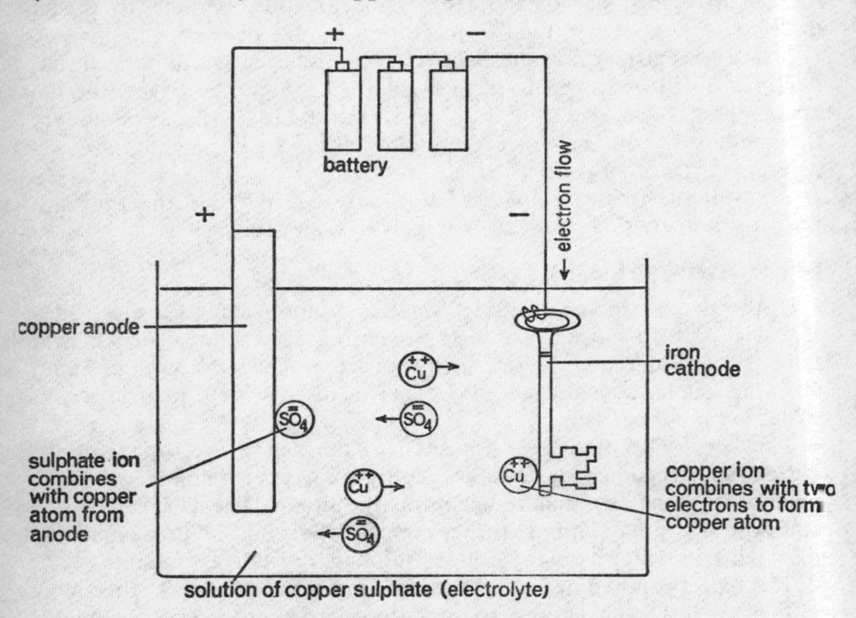

Fig. 44. Electroplating process using copper

terminal. Each sulphate ion, on reaching the copper anode, combines with an atom of copper

$$SO_4^{--} + Cu \longrightarrow CuSO_4 + 2e^-$$

to form another molecule of copper sulphate. The two electrons released in the reaction are absorbed into the anode in an attempt to cancel the deficit of electrons existing there: from the anode they eventually complete their journey back to the battery.

The net result of these reactions is that one copper atom is removed from the anode, and another copper atom is deposited on the cathode. Hence the anode is gradually eaten away while the cathode becomes more and more thickly coated with copper. There is no change in the copper sulphate concentration, since, although one molecule was used up initially, another molecule is formed at the anode.

As we have seen, copper can be deposited on iron. It is not possible, however, to deposit iron on copper, since iron is the more reactive of these two metals.

Non-metals can be electroplated if they are first given a coating of graphite (carbon) so that they will conduct electricity. If the non-metal is stripped away after electroplating an exact impression in metal is left behind. This process is used in the manufacture of gramophone records.

LAWS OF ELECTROLYSIS

In 1833, after carefully carrying out experiments on electrolysis, Michael Faraday published two results which have come to be known as Faraday's laws of electrolysis. The first law can be stated as follows:

The weight of material deposited (or, in the case of gases, liberated) *during electrolysis is directly proportional to the current and to the time for which it operates.*

Thus, doubling the current doubles the weight of material deposited on the cathode, and so does doubling the period of the electrolysis. Since the weight of deposited material is proportional to the current I and to the time t, it is also proportional to their product It. This product is the quantity of electricity or 'charge' which passes, and if the current is measured in amps and the time is measured in seconds the product It gives the charge in 'coulombs'. One coulomb is the charge carried by a current of one ampere in one second.

The second law can be stated as follows:

The weight in grams of any chemical element deposited by 96,500 *coulombs of charge equals the chemical equivalent weight of that element.*

The equivalent weight of an element is defined as the number of grams of it which will combine with or displace 1 g of hydrogen. It is equal to the atomic weight of the element divided by the valency. Fig. 45 shows three electrolysis experiments connected in series. The first contains copper electrodes and copper sulphate solution as electrolyte. The second contains silver electrodes and silver nitrate solution as electrolyte. The third contains aluminium electrodes (fortunately we are not concerned with the overwhelming practical difficulty associated with this particular experiment) and aluminium nitrate solution as electrolyte.

The situation shown in Fig. 45 occurs at the instant when 96,500 coulombs of charge have passed around the circuit. If the current is 0·965 amps this will be after 100,000 seconds (rather more than 27 hours). The atomic weight of copper is given in chemical tables as 63·5 and the valency is +2. Hence the chemical equivalent weight of copper is 31·75, and this is found to be the number of grams of copper deposited in the experiment.

The atomic weight of silver is given as 107·9 and the valency is +1. Hence the chemical equivalent weight is 107·9, which agrees with the weight of silver deposited. Finally, the atomic weight of aluminium is given as 27 and the valency is +3. The chemical equivalent weight is therefore 9, and this, too, is confirmed by the weight of aluminium deposited in the experiment.

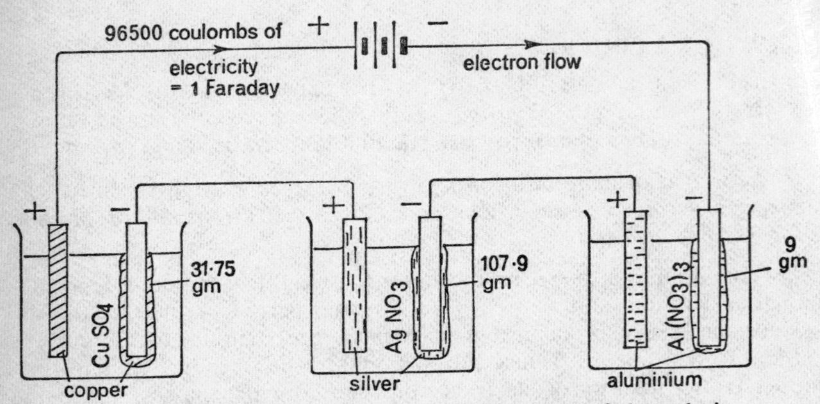

Fig. 45. Demonstration of Faraday's second law of electrolysis

EXAMPLE: How much zinc (atomic weight 65·38, valency = 2) is deposited by a current of 12 A passing through a solution of zinc nitrate for 20 min?

$$\text{Equivalent weight} = \frac{\text{Atomic weight}}{\text{Valency}}$$

$$= \frac{65{\cdot}38}{2}$$

$$= 32{\cdot}69$$

$$\text{Quantity of charge} = \text{Current} \times \text{Time in seconds}$$

$$= 12 \times 20 \times 60$$

$$= 14{,}400 \text{ coulombs}$$

According to the second law of electrolysis,

96,500 coulombs of charge deposits 32·69 g of zinc, or

$$1 \text{ coulomb deposits } \frac{32{\cdot}69}{96{,}500} \text{ g.}$$

Hence 14,400 coulombs deposits

$$\frac{32{\cdot}69 \times 14{,}400}{96{,}500} = 4{\cdot}88 \text{ g of zinc} \quad \textit{Answer}$$

BATTERIES AND CELLS

The familiar word 'battery' is really an abbreviation for a 'battery of cells'. The ordinary dry cell has an e.m.f. of about $1\frac{1}{2}$ V. A 3-V 'battery' is simply a pair of $1\frac{1}{2}$-V dry cells connected in series. A $4\frac{1}{2}$-V battery contains three dry cells in series; a $22\frac{1}{2}$-V hearing-aid battery contains 15 dry cells in series, and so on.

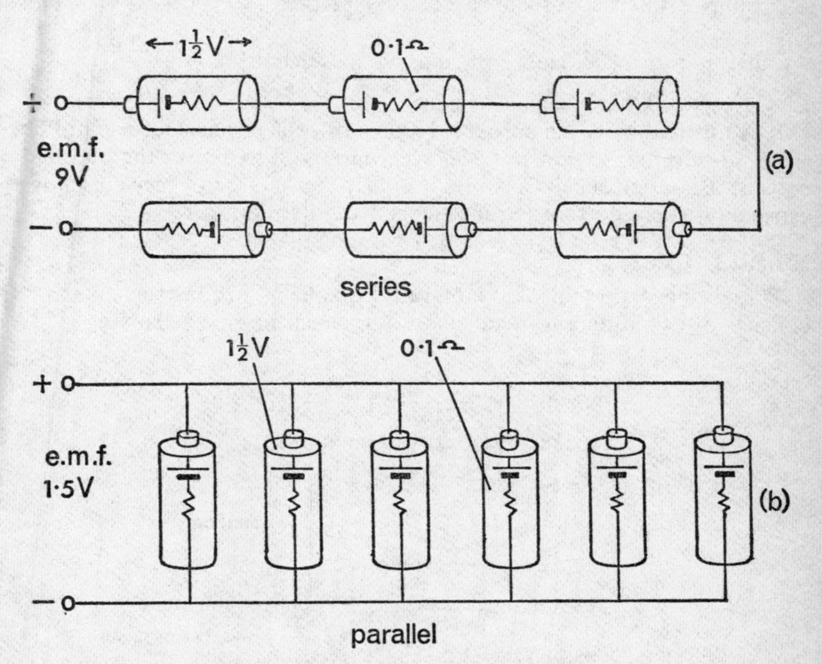

Fig. 46. (*a*) Battery of six cells in series. (*b*) Battery of six cells in parallel

If six dry cells are connected in series the e.m.f. is $6 \times 1\frac{1}{2} = 9$ V. But if the same six cells are connected in parallel the e.m.f. is only $1\frac{1}{2}$ V, the same as for a single cell. In practice, the cells are mostly unlikely to have identical e.m.f.s and will discharge through each other with alacrity; but in theory at any rate there is an advantage to be gained from the parallel connexion: the six cells in parallel can deliver six times as much *current* as each individual cell (or, for that matter, six times as much current as the same six cells connected in series).

The explanation is simply that each cell has some 'internal resistance' (see p. 13) and, in the series connexion, the internal resistances are added together just as the e.m.f.s are added together. Fig. 46 (*a*) shows six cells, each of e.m.f. $1\frac{1}{2}$ V and internal resistance 0·1 Ω, connected in series. Total internal resistance is 0·6 Ω; total e.m.f. is 9 V. Hence the maximum current $\left(=\frac{\text{total e.m.f.}}{\text{total resistance}}\right)$ is

$$9/0{\cdot}6 = 15 \text{ amps}$$

Connecting the same cells in parallel, as shown in Fig. 55 (*b*), gives a combined internal resistance of $\frac{1}{60}$ Ω $\left(\text{from } \frac{1}{R} = \frac{1}{R_1} + \frac{1}{R_2} + \right)$ and a total e.m.f. of $1\frac{1}{2}$ V. Hence the maximum current $\left(=\frac{\text{total e.m.f.}}{\text{total resistance}}\right)$ is $1\frac{1}{2} \div \frac{1}{60} = 90$ amps.

It is unusual to come across a battery of cells connected in parallel, since it is cheaper to construct one very large cell to deliver the required current. Batteries of cells in series, on the other hand, are very common, especially where current requirements are only modest.

SIMPLE CELL

The simplest possible kind of cell capable of producing an e.m.f. consists of two different metal plates immersed in an electrolyte.

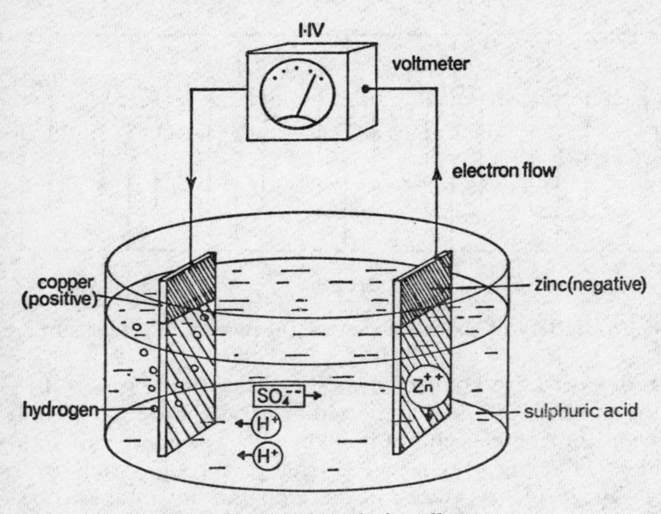

Fig. 47. Simple voltaic cell

The metals we have chosen by way of example in Fig. 47 are zinc and copper, the electrolyte is dilute sulphuric acid. As we shall see, this cell is not very satisfactory, but it does provide an e.m.f. of about 1 V and it will light a torch bulb for a short time.

Like all acids (and bases and salts), the molecules of sulphuric acid tend to break up or dissociate into charged ions: in this case positive hydrogen ions (H^+) and doubly charged negative sulphate ions (SO_4^{--}). This happens whether the cell is connected or not.

Zinc, being a reactive metal, dissolves in the acid, forming positive zinc ions:

$$Zn \longrightarrow Zn^{++} + 2e^-$$

and releasing two electrons per atom.

This process does not go on indefinitely; it stops when the released electrons have accumulated in such numbers on the zinc plate that the force of attraction between them and the positively charged zinc ions prevents any further zinc ions escaping into the solution.

As we might expect, a similar state of affairs exists at the copper plate. But copper is far less reactive than zinc and does not dissolve to anything like the same extent.

When the two plates are connected by a wire the accumulation of electrons on the zinc is given an escape route, and the zinc ions are then able to escape into the solution.

Already present in the electrolyte are hydrogen ions and sulphate ions from the sulphuric acid. The zinc ions (Zn^{++}) team up with the sulphate ions (SO_4^{--}) so that the net result is the displacement of hydrogen ions from the sulphuric acid by the zinc. The displaced hydrogen ions drift across the cell to the copper plate, where each pair of ions steals two electrons to become a neutral molecule of hydrogen:

$$2H + 2e^- \longrightarrow H_2$$

Bubbles of hydrogen gas can be seen on the copper plate while the copper and zinc plates are connected by an external circuit.

Since the copper plate loses electrons to the hydrogen ions, it becomes deficient in electrons, and this encourages more electrons to flow via the conducting circuit from the zinc plate.

Hence there is a current in the external circuit carried by electrons flowing from zinc to copper, and a current in the cell carried by hydrogen ions flowing from zinc to copper.

Summarizing the action of a simple zinc–copper cell, we may write the chemical reaction as

$$Zn + H_2SO_4 \longrightarrow ZnSO_4 + H_2 + \textit{energy}$$

and this conversion of chemical energy into electrical energy is the essential characteristic of every type of cell. When the e.m.f. of a copper–zinc cell is measured it is found to be 1·1 V no matter what electrolyte is used. Each metal has what is known as an *electrode potential* which is a measure of the amount of chemical work performed in getting its ions into solution. Hydrogen, although not a metal, is arbitrarily given an electrode potential of 0 V as a standard of comparison for other electrodes. Metals which develop an electrode potential *more* negative than that of hydrogen are said to have a 'negative potential', while metals which develop an electrode potential *less* negative than hydrogen are said to have a 'positive potential'. In this way all conductors can be arranged in order of their electrode potentials, as shown in Table 4.

Table 4. ELECTRODE POTENTIALS OF SOME COMMON METALS

Element	*Symbol*	*Electrode potential (volts)*
Sodium	Na	−2·71
Aluminium	Al	−1·67
Zinc	Zn	−0·76
Iron	Fe	−0·44
Nickel	Ni	−0·25
Lead	Pb	−0·13
Hydrogen	H	0·00
Copper	Cu	+0·34
Silver	Ag	+0·80
Mercury	Hg	+0·85
Gold	Au	+1·68

Table 4 enables the e.m.f. of any cell to be predicted, since the e.m.f. is the *difference* between the electrode potentials of the two plates. In the simple cell discussed above the plates are copper (potential + 0·34 V) and zinc (potential − 0·76 V). The e.m.f. is therefore 0·34 − (−0·76) = 1·1 V. The table also gives an indication of the chemical activity of the elements listed. Those near the top, such as sodium, are the most active; those near the bottom are relatively inactive or inert.

POLARIZATION

It is found in practice that the e.m.f. of a simple copper–zinc cell very soon falls short of the theoretical 1·1 V. This is because the coating of hydrogen on the copper plate turns it into a hydrogen electrode (potential 0 V) rather than a copper electrode (potential 0·34 V). The cell becomes in effect a zinc–hydrogen cell, for which the theoretical e.m.f. is only 0·76 V, and moreover, the internal resistance is considerably worsened by the layer of hydrogen bubbles. Thus the coating of the copper plate with hydrogen is a serious defect; it is known as *polarization.*

In practical cells polarization is overcome by the addition of a chemical which combines with hydrogen to form water. Such a chemical is called a depolarizing agent or just depolarizer; a chemist would call it an oxidizing agent.

THE DRY CELL

The so-called dry cell is not really dry: its electrolyte is a moist paste instead of a liquid solution. The electrolyte is in fact a salt called ammonium chloride (also known as sal ammoniac), which is in many ways similar to sodium chloride (common salt). A zinc cylinder serves both as the negative plate and as a container for the other ingredients; it is lined with a kind of blotting paper holding a paste of ammonium chloride and water thickened with a little plaster of Paris. Most of the space within the cell is taken up by a paste of manganese dioxide saturated with a solution of ammonium chloride and mixed with powdered carbon to reduce the resistance (Fig. 48).

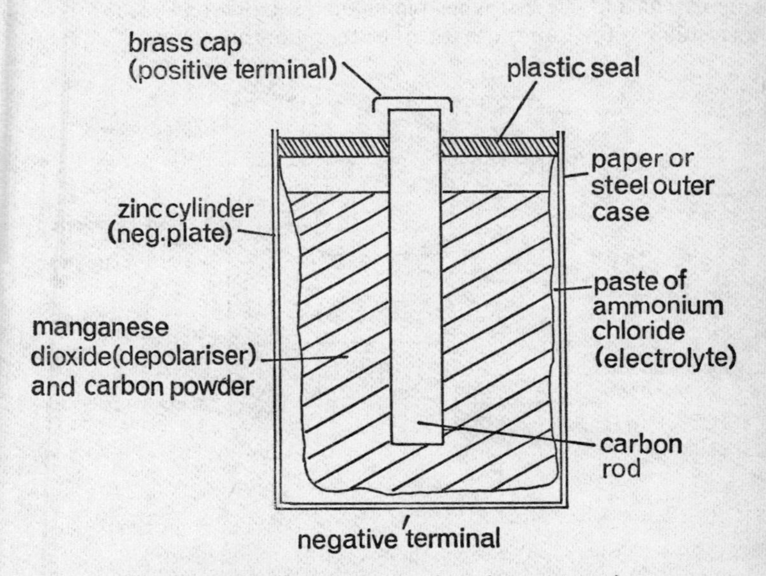

Fig. 48. Dry cell cut away to show its construction

The carbon rod serves merely as a terminal providing an electrical connexion with the manganese dioxide packed around it. Carbon is chemically inert and does not enter into the reactions occurring in the cell; the true positive plate is the manganese dioxide (MnO_2).

The reaction at the positive plate is the conversion of manganese dioxide to manganese sesquioxide (Mn_2O_3), a type of reaction which chemists call a reduction:

$$2MnO_2 + H_2O + 2e^- \longrightarrow Mn_2O_3 + 2(OH^-)$$

and the absorption of two electrons. The resulting hydroxyl (OH^-) ions are largely removed by combination with ammonium ions, and this is the reason for choosing an ammonium salt as the electrolyte.

The reaction at the negative plate is very similar to that of the simple cell described above. Zinc is gradually dissolved by the electrolyte to form zinc (Zn^{++}) ions, while two electrons are released per atom:

$$Zn \longrightarrow Zn^{++} + 2e^{-}$$

When a dry cell is fresh its e.m.f. is about 1·5 V, but because of the production of insoluble manganese compounds, the internal resistance increases rapidly as the cell is used, and the terminal voltage decreases. If the cell is left to rest, however, the manganese dioxide recovers to some extent and the cell is usable again. For this reason, dry cells are more suitable for intermittent rather than continuous use.

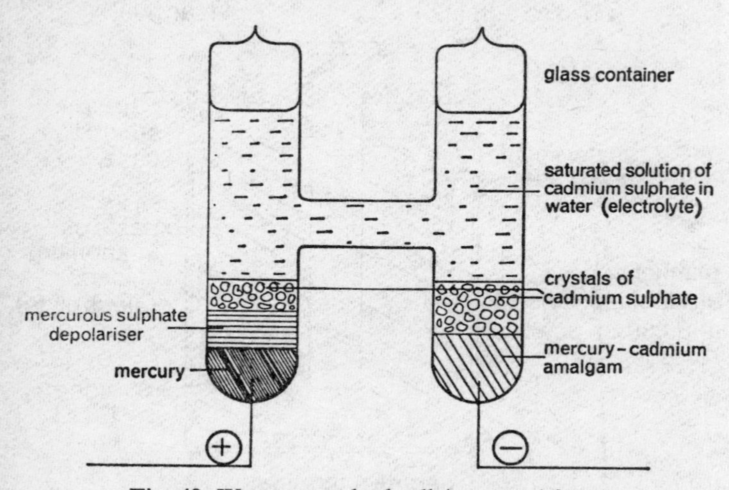

Fig. 49. Weston standard cell (saturated form)

A less bulky dry cell is the so-called *mercury battery* used in hearing aids, exposure meters, and some automatic cameras. Its big advantage over the zinc–carbon dry cell is that its e.m.f. of 1·35 V remains reasonably steady throughout its life. The negative plate is zinc, the positive plate is mercury in contact with mercuric oxide, and the electrolyte is potassium hydroxide mixed with potassium zincate.

The Weston cell (Fig. 49) is not often encountered outside the laboratory, but it is important, since its e.m.f. is accurately known to be 1·0183 V at 20° C, and the cell therefore acts as a standard against which other voltages can be checked.

STORAGE BATTERIES

All the cells mentioned so far have to be thrown away when the chemical energy is exhausted. Cells which can be recharged, i.e. the chemical energy can be restored, are called *accumulators* or *storage batteries*. (These are sometimes referred to as 'secondary cells' to distinguish them from the ordinary throw-away 'primary cells'. This nomenclature arose in the days when 'primary' cells were the only available source of current for recharging accumulators.)

By far the most common form of storage battery is the *lead–acid* type

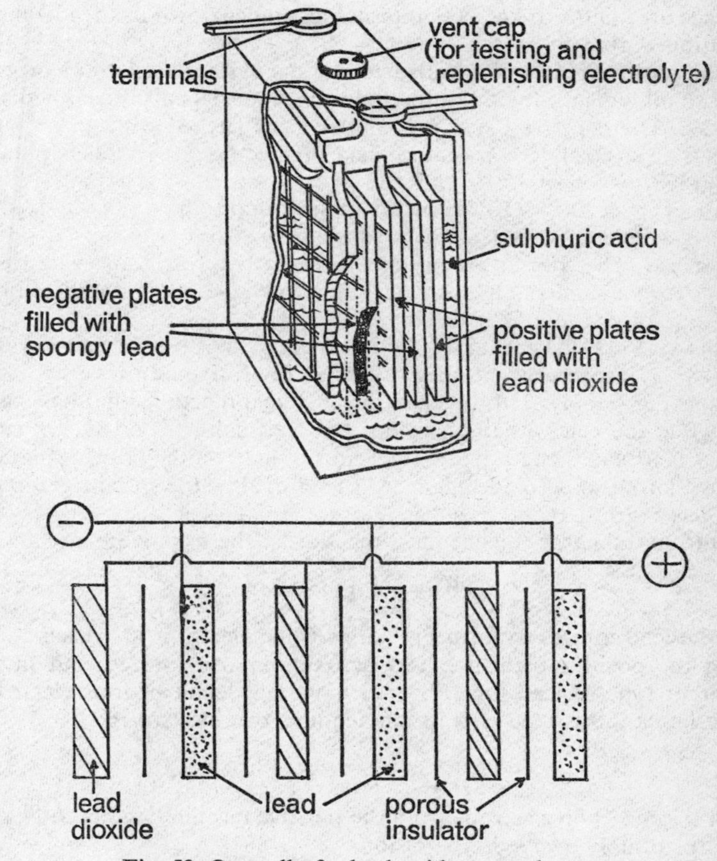

Fig. 50. One cell of a lead–acid storage battery

found in every motor vehicle. It has a very low internal resistance (about 0·01 Ω), and can therefore deliver the very high currents (200 A or more) needed by the starter motor. Its e.m.f. is approximately 2 V per cell.

In order to achieve the greatest possible surface area of the plates, each cell contains several positive plates and a similar number of negative plates separated by porous insulators, as shown in Fig. 50. The surface area of the plates is very important, since it is on the interface between the plate and the electrolyte that the essential reaction of any cell takes place, namely the transfer of ions. Hence the greater the area of the plates, the greater is the amount of chemical action and, in this case, the greater the amount of energy stored. Naturally, the large surface area helps to keep the internal resistance down to the low figure mentioned above.

The positive plates of a lead–acid storage battery are lattices or grids of lead alloy packed with a brown porous material called lead peroxide (PbO_2). The negative plates are similar grids packed with grey spongy lead. The electrolyte is sulphuric acid of specific gravity (density compared with water) 1·25.

When the cell runs down, each plate is found to have a white coating of lead sulphate ($PbSO_4$) and the specific gravity of the acid drops to 1·2 or less. The specific gravity of the electrolyte (easily measured with an instrument called a hydrometer) therefore gives a good indication of the state of charge of the cell.

The cell may be recharged by connecting the positive and negative terminals, respectively, to the positive and negative terminals of a low-voltage d.c. supply. This reverses the chemical reactions that have gone on while the cell was discharging; the lead sulphate on the negative plates is turned into lead, and the lead sulphate on the positive plates is turned into lead peroxide again. At the same time the specific gravity of the acid rises to 1·25.

During discharge the chemical reaction at the negative plate is

$$Pb \longrightarrow Pb^{++} + 2e^-$$

i.e. the lead ionizes to form Pb^{++} ions while electrons are released.

At the positive plate the lead peroxide is also ionized, but into a different type of lead ion (Pb^{++++}) having a deficit of four electrons. When it can, this type of lead ion acquires two electrons:

$$Pb^{++++} + 2e^- \longrightarrow Pb^{++}$$

So it removes two electrons from the positive terminal and is converted into the doubly charged type of ion.

Hence at both positive and negative plates there are doubly charged lead ions (Pb^{++}) which displace hydrogen from the sulphuric acid to form lead sulphate. The oxygen part of the lead peroxide reacts with hydrogen ions to form water, the net result being the formation of one water molecule for each acid molecule broken up. The acid therefore becomes diluted while the cell is working.

Exactly the reverse reactions take place when the cell is recharged. The reactions are conveniently summarized by the formula

Lead peroxide		Lead		Sulphuric acid		Lead sulphate		Water
PbO_2	+	Pb	+	$2H_2SO_4$	$\rightleftharpoons$	$2PbSO_4$	+	$2H_2O$
Positive plate		Negative plate		Electrolyte				

reading from left to right for the discharge action, and from right to left for the recharging action.

The *nickel–iron alkali cell* (known in America as the Edison cell) is much more robust than the lead–acid cell. It weighs less, is mechanically rugged, is undamaged by overloading or overcharging, and has a far longer life. But, on the other hand, it is appreciably more expensive than the lead–acid type. Its e.m.f. is about 1·25 V.

The positive plate consists of a steel grid into which pencil-shaped perforated steel tubes filled with nickel hydroxide are inserted. The negative plate is a similar steel grid containing iron oxide. A solution of 21 per cent caustic potash in water acts as the electrolyte. This solution is unaffected by the charging or discharging actions: it simply serves to transfer ions from one plate to the other.

The *nickel–cadmium* storage cell is a development of the nickel–iron cell. Its positive plates contain nickel hydroxide, and its negative plates contain cadmium. Again the electrolyte is a solution of caustic potash in water. Small leakproof versions of this cell are used in a variety of portable equipment, such as the photographer's electronic flashgun.

An even smaller accumulator is the *silver–zinc* cell. The positive plate is actually silver oxide, and the negative plate is a sheet of zinc. Once again the electrolyte is a solution of caustic potash. This cell has, for its weight, a high capacity for storing energy. It will also tolerate a large amount of misuse. Its e.m.f. is 1·5 V and remains fairly constant.

The two big disadvantages of storage cells are their weight and their cost. Scientists are at present desperately trying to find a lightweight inexpensive battery that will make the electric car a practicable proposition. The best hopes seem to lie in what is known as the 'fuel cell'.

In a petrol engine the fuel is burned (i.e. combines with oxygen) to produce heat, which is in turn converted with considerable losses into mechanical energy. In a *fuel cell* the chemical energy of the fuel is converted directly into electrical energy.

The simplest fuel cell is rather like the reverse of the electrolysis of water. Hydrogen (or a cheaper fuel) and oxygen (or air) are passed through porous electrodes into an electrolyte. As the gases dissolve, an e.m.f. develops between the electrodes, as we saw in the case of zinc dissolving in the simple cell (p. 81). The great merit of this arrangement is that relatively small quantities of fuel will maintain the e.m.f. for a very long time.

SUMMARY

Electrolysis is the breaking up of a chemical compound by means of an electric current. Solutions of acids, bases, and salts are called electrolytes; they can conduct electricity because they contain charged ions.

In electrolysis the negatively charged electrode is called the cathode and the positively charged electrode is the anode. In the electroplating process material dissolved from the anode is deposited on the cathode.

Faraday's laws of electrolysis state: (i) the weight of any material liberated by electrolysis is directly proportional to the quantity of charge passing (i.e. to the product of current × time); (ii) 96,500 coulombs of charge liberates the equivalent weight (i.e. atomic weight ÷ valency) in grams of any chemical element.

If *n* identical cells are connected in series the total e.m.f. is *n* times that of a single cell; if they are connected in parallel the total e.m.f. is equal to that of a single cell, but the combination can deliver a current *n* times as great as that from a single cell.

All cells convert chemical energy into electrical energy. In the simple cell electrical energy is produced when zinc dissolves in sulphuric acid. The e.m.f. of a cell can be predicted from the difference between the electrode potentials of its two plates.

The simple cell suffers from polarization, i.e. the accumulation of hydrogen on its positive plate; this is overcome in practical cells by the addition of a depolarizing chemical which converts hydrogen to water.

Storage batteries can be recharged when their chemical energy is exhausted. This is achieved by passing a current through the battery to reverse the reactions that have taken place during discharge.

CHAPTER SIX

ELECTROMAGNETIC FORCES I: THE MOTOR EFFECT

From what was said in Chapter Four about the magnetic field associated with every current-carrying conductor, there arises the question of what happens when a wire carrying a current is placed in an external magnetic field. As might be expected, the field created by the current interacts with the external field, and as a result, a force is exerted on the wire—just as there is a force between two permanent magnets when their fields interact. It should be understood that the force is exerted on the *current* rather than on the conductor: a copper wire, for example, is not affected by a magnet unless it is carrying a current. More important, it is possible to have a current without a conductor (such as a stream of electrons passing through the empty space in a cathode-ray tube), and this current can be deflected magnetically. Incidentally, the rapid repetitive motion of the electron beam scanning a television screen is produced by magnetic deflecting coils around the neck of the picture tube; the image in a television camera is scanned in the same way.

Before discussing the effects of the force experienced by a conductor in a magnetic field, let us investigate its magnitude.

The factors which govern the force acting on a current-carrying wire in a magnetic field are: (i) the length of the wire; (ii) the size of the current; and (iii) the flux density of the field. If a straight wire of length l metres carrying a current I amps and passing at right angles to a magnetic field of flux density B webers per square metre experiences a force of F newtons the relationship between B, I, l, and F is simply

$$F = BIl \qquad (1)$$

The simplicity of this extremely important relationship is not merely fortuitous, it is deliberately contrived by our choice of the weber per square metre as the unit of flux density. Since $B = \mu H$, we can write eqn. (1) in the alternative form

$$F = \mu HIl \text{ newtons} \qquad (1a)$$

where H is the field strength in ampere-turns per metre, and $\mu = \mu_r\mu_0$ is the permeability of the medium.

EXAMPLE: Two very long parallel straight wires, each carrying a current of 1 amp, are placed 1 metre apart in a vacuum. What is the force between them for each metre of their length?

The intensity of the magnetic field at a point r metres from a long straight wire is given by eqn. (3) on p. 56. It is

$$H = \frac{I}{2\pi r}$$

where I is the current in amps carried by the wire. Thus the magnetic field around the first of the parallel wires due to the current in the second is, since $I = 1$ amp and $r = 1$ m,

$$H = \frac{1}{2\pi \times 1}$$
$$= \frac{1}{2\pi} \text{ At/m}$$

Applying the formula for the force exerted on a current-carrying conductor (eqn. (1a)) to 1 metre of the first wire:

$$\mu = 4\pi \times 10^{-7} \text{ for vacuum}$$
$$H = \frac{1}{2\pi} \text{ At/m}$$
$$I = 1 \text{ amp}$$
$$l = 1 \text{ metre}$$
$$F = \mu H I l \text{ newtons}$$
$$= 4\pi \times 10^{-7} \times \frac{1}{2\pi} \times 1$$
$$= 2 \times 10^{-7} \text{ newtons} \quad \textit{Answer}$$

Thus, when a current of exactly 1 amp flows in each of two long parallel wires 1 metre apart in a vacuum the force between the wires is 2×10^{-7} newtons per metre.

The above example is really a definition of the ampere.

In the SI unit system the ampere is defined internationally as *the current in a long straight wire in a vacuum which exerts a force of exactly 2×10^{-7} newtons per metre on a parallel long straight wire 1 metre away carrying the same current.*

Once the ampere has been defined the other electrical units can be derived from it. Thus the SI definition of the volt is the *potential difference between two points on a conducting wire when a current of 1 amp passing between them dissipates a power of 1 watt* (*i.e. 1 joule of energy per second*).

The ohm is defined internationally as *the resistance between two points on a conducting wire when a potential difference of 1 volt applied to these points produces a current of 1 amp.*

Table 5. SI UNITS AND THEIR C.G.S. EQUIVALENTS

Quantity	*Symbol*	*SI unit*	*C.G.S. equivalent*
Length	l	metre, m	100 cm
Mass	m	kilogram, kg	1,000 g
Time	t	second, s	1 sec
Force	F	newton, N	10^5 dynes
Work or energy	W	joule, J	10^7 ergs
Power	P	watt, W	10^7 ergs/sec
Charge	Q	coulomb, C	0·1 unit
Current	I	ampere, A	0·1 unit
E.M.F. (or p.d.)	E (or V)	volt, V	10^8 units
Resistance	R	ohm, Ω	10^9 units
Inductance	L	henry, H	10^9 units
Capacitance	C	farad, F	10^{-9} units
Magnetomotive force	F	ampere-turn, At	$0{\cdot}4\pi$ gilberts
Magnetic field strength	H	ampere-turn per metre, At/m	$4\pi \times 10^{-3}$ oersted
Magnetic flux	ϕ	weber, Wb	10^8 maxwells
Flux density	B	weber per sq. metre, Wb/m^2 (or tesla)	10^4 gauss
Permeability of free space	μ_0	$4\pi \times 10^{-7}$ units	1 unit

DIRECTION OF MAGNETIC FORCE

Force is a *vector* quantity, i.e. it has *direction* as well as magnitude. As shown in Fig. 51 (*a*), the direction of the force experienced by a conductor in a magnetic field is at right angles to the field and at right angles to the (conventional) current. If two of these directions are known the third can be predicted by a simple stratagem known as Fleming's left-hand rule (Fig. 51 (*b*)).

To predict the direction of the force on a conductor, the thumb, first finger, and second finger of the left hand are extended at right angles to one another, with the first finger pointing in the direction of the field (N to S) and the second finger pointing in the direction of the current. The thumb now indicates the direction of the force on the conductor, and hence the direction in which it tends to move. Applying the left-hand rule to parallel conductors (Fig. 52), it is found that when the currents are in the same direction the conductors are attracted to each other; when the currents are in opposite directions the conductors repel each other. The effect of attraction between wires carrying current in the same direction is in fact observed on large coils in power plants. Under short-circuit conditions adjacent turns of these coils can be drawn together with such force that the windings are damaged.

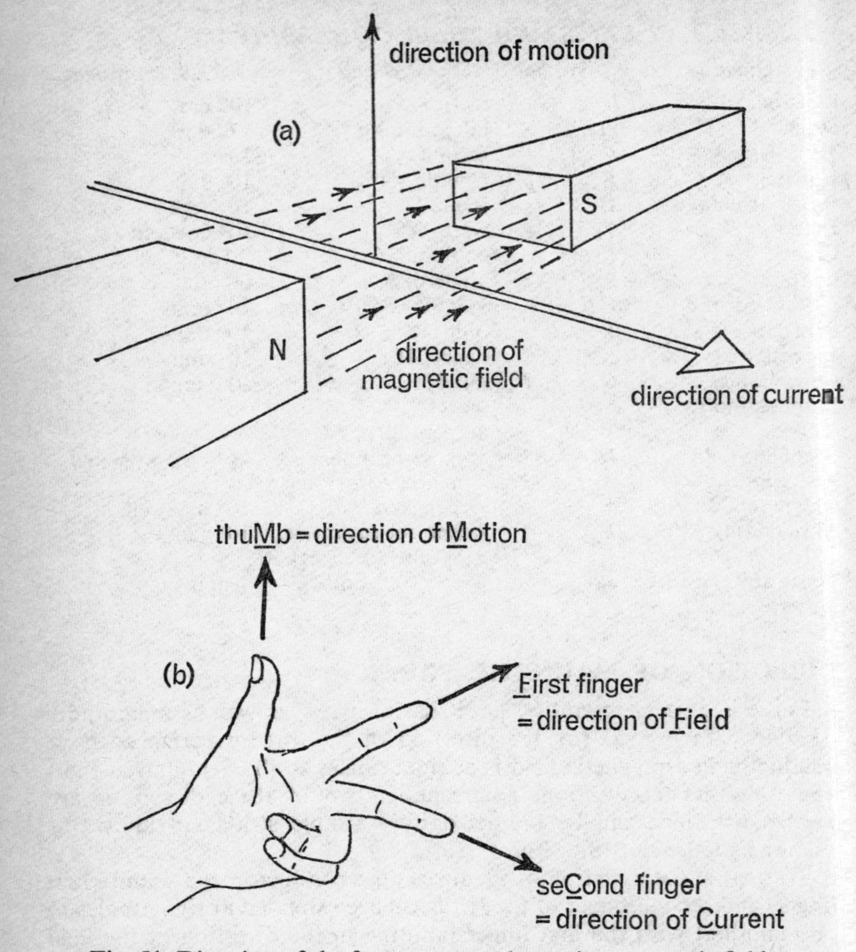

Fig. 51. Direction of the force on a conductor in a magnetic field

MOVING-COIL GALVANOMETER

The force on a conductor in a magnetic field is the basis of most electrical measuring instruments, including ammeters and voltmeters. These instruments are all modifications of the *moving-coil galvanometer*, illustrated in Fig. 53.

The moving-coil galvanometer consists of a flat rectangular coil of fine wire pivoted between the poles of a permanent magnet. A soft-iron core (which is *not* free to move) fixed inside the coil, together with the concave shape of the magnet's pole pieces, ensures that the plane of the

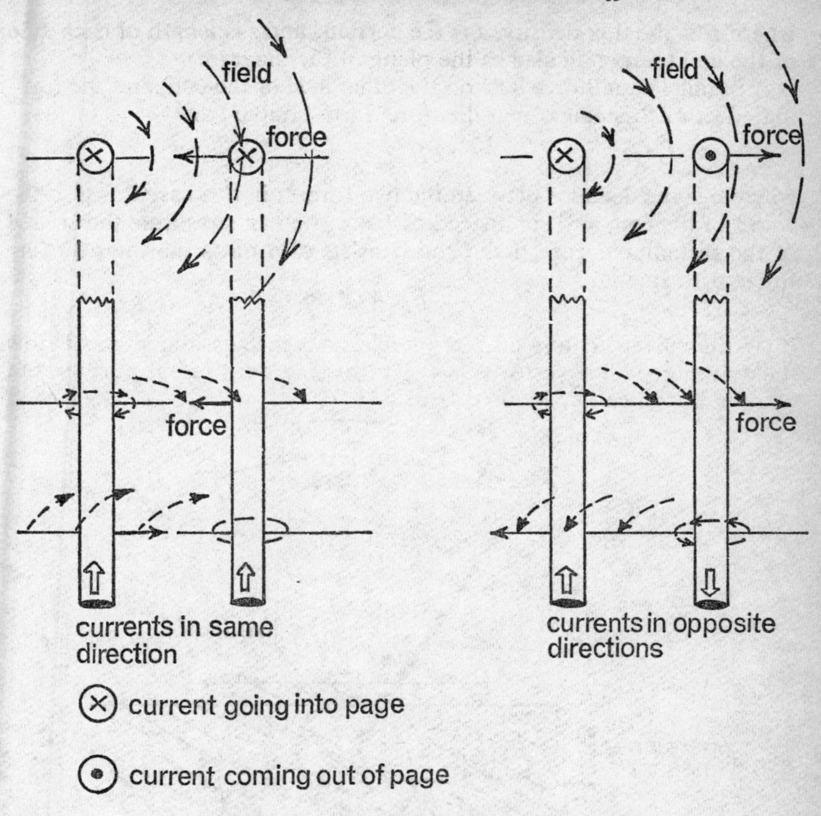

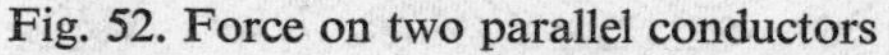
Fig. 52. Force on two parallel conductors

coil is always parallel to the magnetic field. Current enters and leaves the coil through hair springs which also serve to resist the rotation of the coil and return it to its original position.

Applying the left-hand rule to the *sides* of the coil in Fig. 53, we find that on the right, where current flows into the plane of the diagram, the motion is downward. On the left of the coil, where current comes out of the plane of the diagram, the motion is upward. Hence the coil rotates clockwise and the pointer moves from left to right over the scale.

As the coil rotates, the hair springs are wound up until the turning force they exert is equal and opposite to the magnetic force causing the rotation. Force on each conductor at right angles to the field is

$$F = BIl \text{ newtons (see p. 89)}$$

where B is the flux density, I is the current, and l is length of each side of the coil (perpendicular to the plane of the diagram).

A parallel equal force acts on the other side of the coil, and the turning effect of this pair of parallel forces (or 'couple') is

$$BIl \times a$$

where a is the distance between the two forces: in this case it is also the width of the coil, and so, instead of $l \times a$, we can substitute the area A of the rectangular coil. If the coil consists of n turns the total turning force or torque is

$$BIAn$$

(If the current enters and leaves via hair springs above and below the coil there will have to be $n + \frac{1}{2}$ turns; for our present purpose the extra half turn can be ignored.)

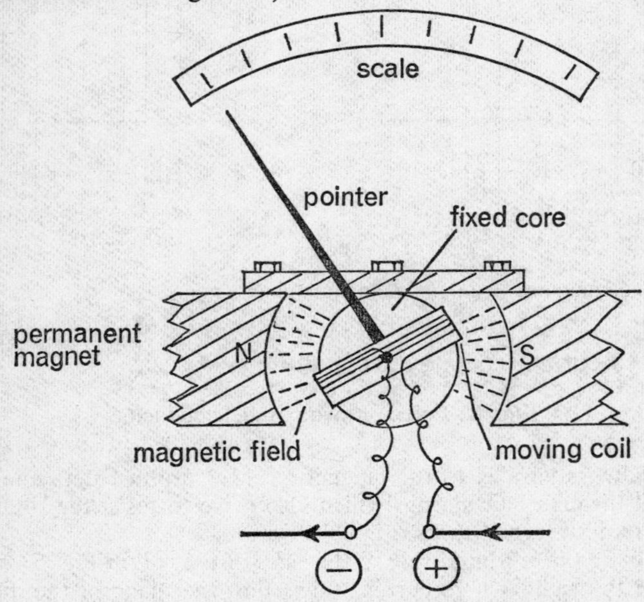

Fig. 53. Action of a moving-coil galvanometer

The coil comes to rest when the torque is balanced by the restoring torque of the springs—which is proportional to the angle θ turned through. Since B, A, and n are fixed for any given instrument, this angle θ is directly proportional to the current I: the larger the current, the bigger the deflection.

A sensitive meter is one in which a small current turns the coil through a large angle. Hence the ratio θ/I is a measure of sensitivity.

Sensitivity (θ/I) depends on B, A, and n, which should therefore be as large as practicable.

The area of the coil A is limited by the physical size of the meter. The number of turns n can be made large if very thin wire is employed, but this increases the resistance of the coil. Hence the best way of achieving a high sensitivity is to use a very intense magnetic field, for which a powerful permanent magnet with small pole pieces is required.

Magnets used in galvanometers are generally made from alloys such as Alnico (51 per cent iron, 24 per cent cobalt, 14 per cent nickel, 8 per cent aluminium, and 3 per cent copper).

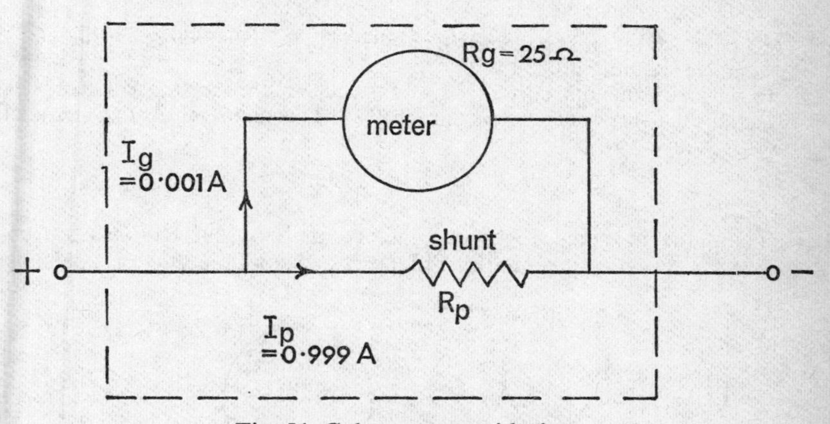

Fig. 54. Galvanometer with shunt

The scale reading of a galvanometer is a measure of the current flowing in the moving coil (since $\theta \propto I$ as explained above). The scale can therefore be marked off in units of current such as milliamps (mA) and microamps (μA).

If a very small current of a few milliamps produces a full-scale deflection a larger current would damage the instrument. However, the galvanometer can easily be converted into an ammeter for measuring much larger currents by means of a 'shunt', i.e. a resistor in parallel with the moving coil. If the resistor is small it acts as a by-pass for most of the current, so that only a tiny fraction (perhaps a few milliamps) of the total actually flows through the meter. The action of the shunt is illustrated by the following example.

EXAMPLE: A moving-coil galvanometer has a resistance of 25 Ω, and shows a full-scale deflection when passing a current of 1 mA (0·001 A). What resistor connected across its terminals will turn the instrument into an ammeter measuring currents up to 1 A?

As shown in Fig. 54, when a total current of 1 A is divided between the meter and the shunt only 0·001 A may pass through the meter. Therefore the current through the shunt is

$$1 \cdot 0 - 0 \cdot 001 = 0 \cdot 999 \text{ A}$$

Applying Ohm's law in the form $V = IR$ to the galvonometer, the p.d. across the meter is 0·001 × 25 (volts). Since the shunt and the meter are connected in parallel, the p.d. across the shunt is also 0·001 × 25 (volts). Applying Ohm's law in the form $R = V/I$ to the shunt:

$$\text{Resistance of shunt} = \frac{0 \cdot 001 \times 25}{0 \cdot 999}$$
$$= 0 \cdot 025025 \ \Omega$$

Thus a 0·025-Ω resistor connected in parallel with the galvanometer turns it into an ammeter reading up to 1 A.

In general, if the total current is I_t and the current through the galvanometer is I_g and the galvanometer resistance is R_g, the required value of the parallel resistor is given by

$$R_p = \frac{I_g \times R_g}{(I_t - I_g)} \text{ ohms}$$

The same moving-coil galvanometer can be converted into a *voltmeter* for measuring potential differences. A moving-coil voltmeter is

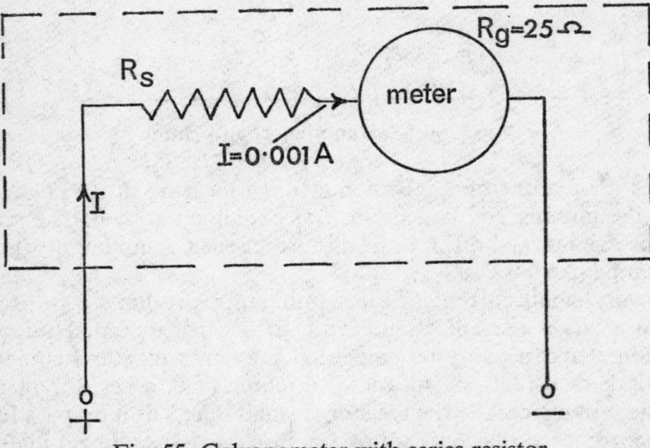

Fig. 55. Galvanometer with series resistor

simply the basic galvanometer with a large resistor connected in series. The action of the series resistor is illustrated by the following example.

EXAMPLE: A moving-coil galvanometer has a resistance of 25 Ω and shows a full-scale deflection when passing a current of 1 mA. What resistor connected in series will turn the instrument into a voltmeter measuring potential differences up to 2 V?

Since the resistor and the meter are in series, the same current (1 mA for full-scale deflection) flows through both, as shown in Fig. 55.

Applying Ohm's law in the form $R = V/I$ to the voltmeter terminals, the *total* resistance is $\frac{2}{0{\cdot}001} = 2{,}000\ \Omega$.

Of this total, the galvanometer accounts for 25 Ω; the remaining 1,975 Ω (i.e. 2,000 — 25) must be contributed by the series resistor. Hence a 1,975-Ω resistor in series with this galvanometer turns it into a voltmeter reading up to 2 V.

In general, if the required full-scale deflection is V volts, while the current which actually produces full-scale deflection is I_g amps and the resistance of the galvanometer is R_g ohms, the value of the series resistor is given by

$$R_s = \frac{V}{I_g} - R_g \text{ ohms}$$

ELECTRIC MOTORS

Suppose the restraining springs were removed from a moving-coil galvanometer; would the coil rotate through a complete circle (assuming that nothing got in its way)?

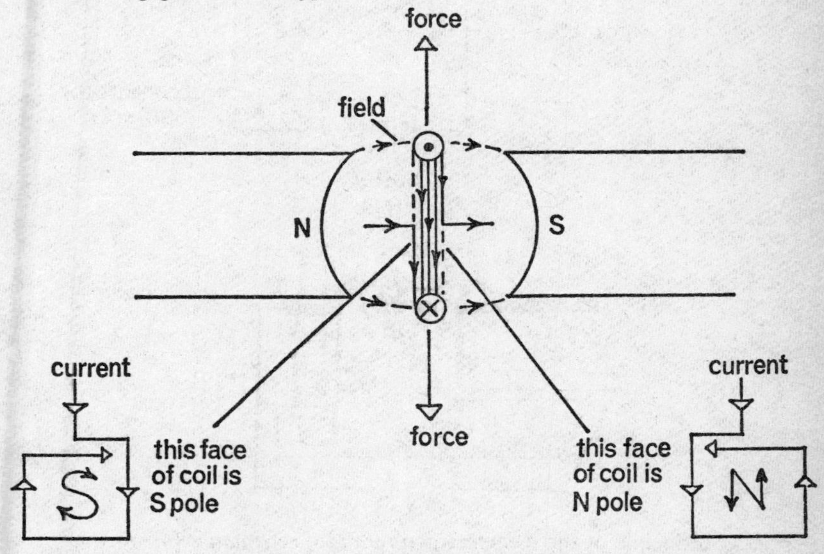

Fig. 56. Coil lying with its plane perpendicular to the field experiences no torque

Unfortunately it would not; for no turning effect is exerted on the coil when it reaches the position shown in Fig. 56.

In this position the N pole of the coil is adjacent to the S pole of the permanent magnet, and the S pole of the coil is adjacent to the N pole of the magnet. Since unlike poles attract, the coil will tend to remain in this position.

Viewing the coil from above, with the plane of the coil lying at right angles to the field, it is clear that the force acting on one side of the coil

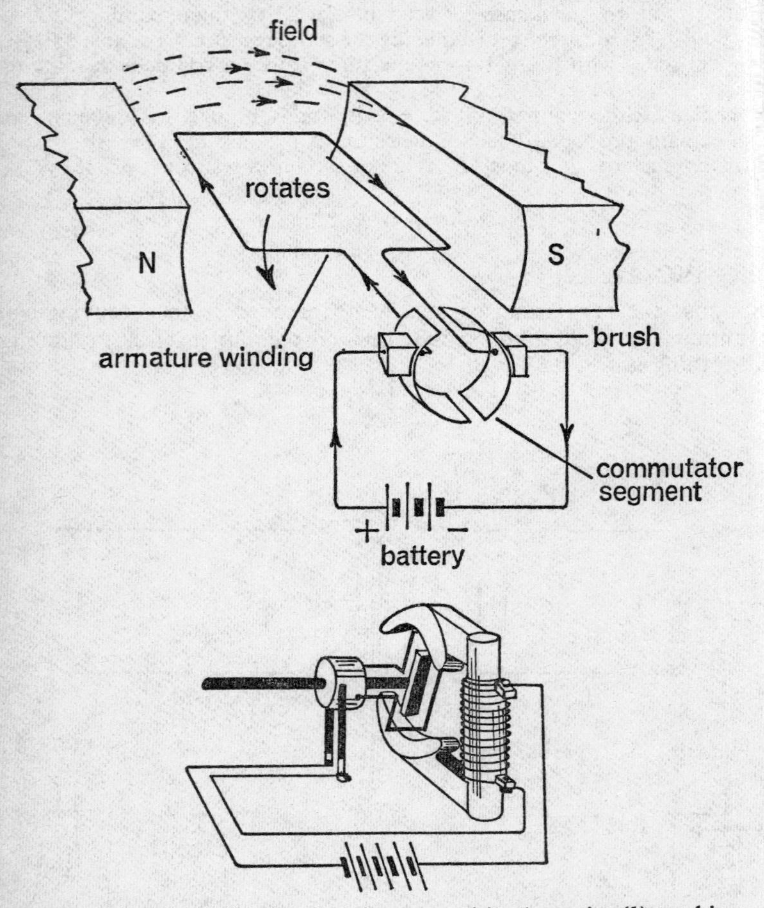

Fig. 57. Principle of the d.c. electric motor: (*a*) schematic; (*b*) working model

is in line with the force acting on the other side: their turning effect is therefore nil.

For this reason the galvanometer cannot be made to turn through more than half a revolution at most, no matter how great the current. To achieve continuous rotation we need some sort of switching device to exchange the magnetic poles on the faces of the coil every half turn; in other words, to reverse the current twice per revolution.

A direct-current motor is essentially a galvanometer movement fitted with an automatic switch called a *commutator* to reverse the current every half turn. A very simple motor is shown in Fig. 57, together with its commutator.

The commutator is a ring split into two segments which are connected to the two ends of the moving coil and rotate with it. Current enters and leaves the coil via 'brushes'—blocks of carbon which slide on the cylindrical commutator segments.

When the coil in Fig. 57 reaches the vertical position the left-hand commutator segment, which had been in contact with the positive brush, comes into contact with the negative brush. At this same instant the right-hand segment, which had been touching the negative brush, makes contact with the positive brush. The direction of the current in the coil is consequently reversed, and the force which had been pulling the coil up is now changed into one pulling the coil down. This happens whenever the plane of the coil is vertical, i.e. twice in each revolution.

The coil, called the armature winding, rotates together with its iron core. The magnetic field is usually provided by an electromagnet as shown in Fig. 57 (*b*). In practice, the armature carries a number of separate windings connected to an equal number of pairs of segments on the commutator.

BACK E.M.F.

There is no real difference between the construction of an electric motor and a generator (see p. 112). In fact, an e.m.f. is always generated in the armature of an electric motor when it is running, and as we shall see in Chapter Seven, this e.m.f. acts like a battery inside the armature, which opposes the applied voltage.

There is no 'induced' current when the armature is at rest. Consequently, the current in the armature at the instant the motor is switched on (i.e. the supply current alone) is much greater than the net armature current when it is turning at full speed (i.e. supply current *less* the induced current).

The sudden surge of current which flows through the armature as soon as the motor is switched on could easily produce enough heat to burn out the windings and their insulation. To avoid this, a special kind of variable resistor is used to regulate the current, allowing it to increase gradually until the motor is running at full speed (Fig. 58).

The force acting on the coil of a galvanometer was shown (p. 94) to have a turning effect proportional to $BIAn$. Not surprisingly, the same expression applies to the torque produced by an electric motor. In a motor, however, the flux is usually provided by an electromagnet or 'field winding', and the flux density B is proportional to the current (I_f) in the field windings. Since the area (A) of the armature windings and the number of turns (n) are both fixed, the torque is proportional to the armature current I_a and to the field current I_f. Hence the torque of a motor is proportional to the product $I_a \times I_f$.

The armature and the field windings draw their respective currents from the same d.c. source. If they are connected in parallel across the supply as shown in Fig. 59 (*a*) the motor is described as 'shunt connected'. If the armature and field windings are connected in series across

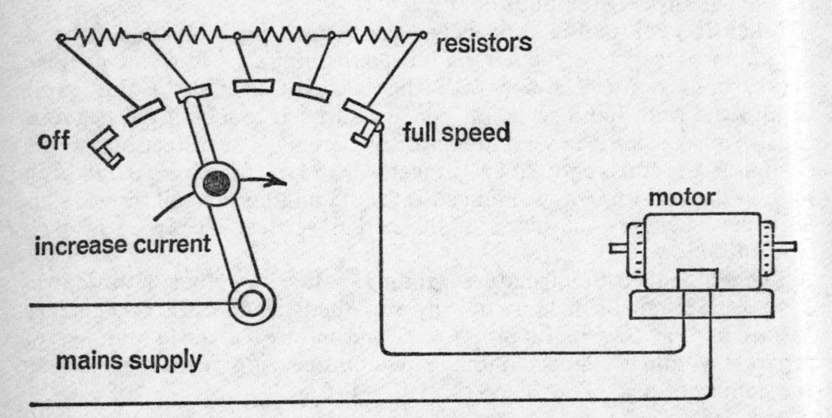

Fig. 58. Starting resistor used to increase current as motor picks up speed

the supply, so that the same current flows in each, as shown in Fig. 59 (*b*), the motor is described as 'series connected'. If they are connected partly in series and partly in parallel, as shown in Fig. 59 (*c*), the motor is described as 'compound connected'.

In the shunt-connected motor the current drawn from the source divides between the armature and field windings. When the load being moved by the motor is increased the armature is slowed down; consequently, the induced current decreases and more current flows in the armature windings. But more current passing through the armature leaves less current to flow through the field windings, i.e. I_a increases and I_f decreases: the product $I_a I_f$ is practically unchanged. Thus shunt-connected motors have a fairly constant torque and are useful in applications calling for steady speeds, e.g. cooling fans, water pumps, etc.

In the series-connected motor the same current flows in both the armature and the field windings. Since $I_a = I_f$, the torque is proportional to the *square* of the current. What happens in this case when the load being moved by the motor is increased? The motor slows down, the induced current decreases, and extra current is drawn from the source.

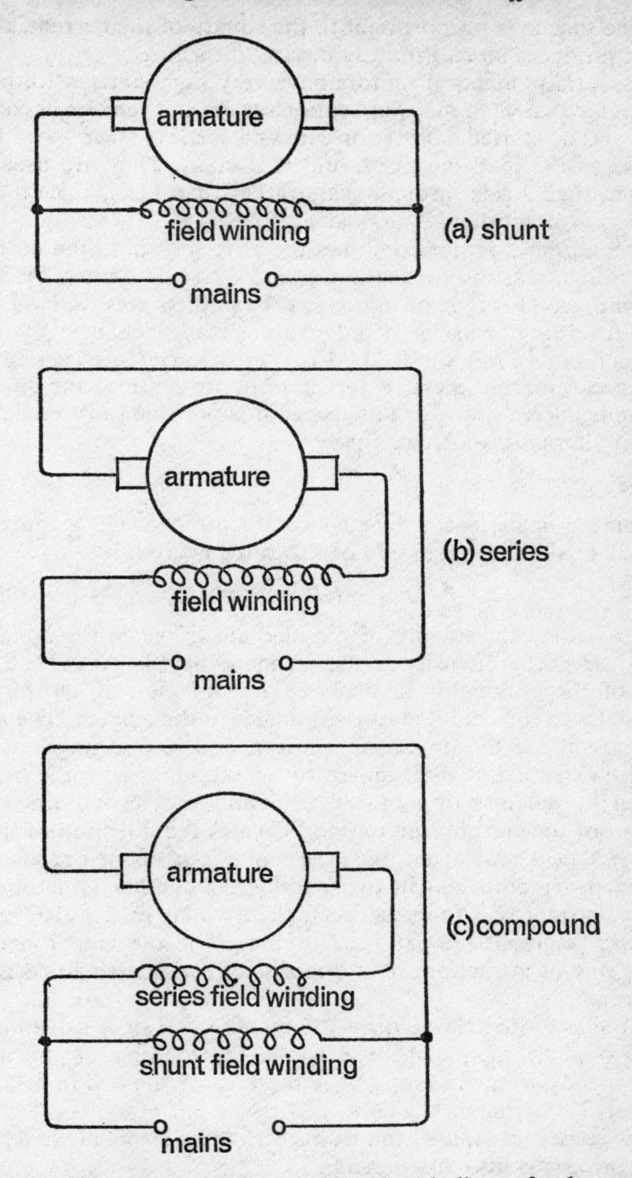

Fig. 59. Three methods of connecting the windings of a d.c. motor

Since the torque is proportional to the square of the current, the extra current produces an enormously increased torque.

Thus, series-connected motors have very high starting torques, and are therefore useful in such applications as electric vehicles, locomotives, hoists, and as starter motors for cars and lorries. Their speed tends to decrease as the load increases, and vice versa. They are usually connected to their loads through gears rather than belts, since they 'race' dangerously if accidentally unloaded.

The compound motor combines the virtues of both the other types. It has a high starting torque and a speed that is limited by its shunt field winding. This type of motor is used in lifts, presses, lathes, etc.

The direction of rotation of a d.c. motor may be changed by reversing the direction of *either* the field current *or* the armature current. If *both* the field current and the armature current are reversed the direction of rotation is unchanged. For this reason it is possible to run a d.c. motor from an alternating-current supply.

SUMMARY

A force is experienced by any conductor carrying a current in a magnetic field. The magnitude of this force is given by

$$F = \mu H I l \text{ newtons}$$

where μ is the permeability of the medium in which the conductor is situated, H is the intensity of the magnetic field in At/m, and l is the length of the conductor in metres. The direction of the force is at right angles to the field and at right angles to the current. The *left-hand rule* states that if the thumb, first finger, and second finger of the left hand are extended at right angles to one another, with the first finger pointing in the direction of the field and the second finger in the direction of the current, the thumb indicates the direction of the force.

The practical unit of current, the ampere, is defined as the current which, when maintained in two parallel conductors 1 metre apart in vacuum, produces a force between them equal to 2×10^{-7} newtons per metre length. If the parallel currents are in the same direction the force is one of attraction; if in opposite directions the force is one of repulsion.

The angle turned through by the moving coil of a galvanometer is proportional to $BIAn$, where B is the flux density of the magnetic field, I is the current in amps, A is the area of the coil in m^2, and n is the number of turns on the coil.

A low resistance (shunt) connected across the terminals of a galvanometer converts it into an ammeter.

A high resistance connected in series with a galvanometer converts it into a voltmeter.

A d.c. motor consists essentially of a rotating armature, a magnetic field produced by a permanent magnet or an electromagnet, a commutator for reversing the direction of the current in the armature windings every half turn, and 'brushes' to supply current to the armature windings through the commutator segments.

The current drawn from the d.c. supply decreases as the motor gathers speed: this is because the moving armature generates a current which opposes the current from the supply.

D.C. motors are classified as series, shunt, or compound, depending on how the armature and field windings are connected together.

CHAPTER SEVEN

ELECTROMAGNETIC FORCES II: THE GENERATOR EFFECT

The development of electrical technology on a commercial scale began when it was freed from its former dependence on cells and batteries. After Oersted had shown (p. 50) that magnetic fields were associated with, and could be produced by, electricity the search began for the opposite effect, i.e. the production of electricity from magnetism.

Michael Faraday, in particular, carried out a series of experiments

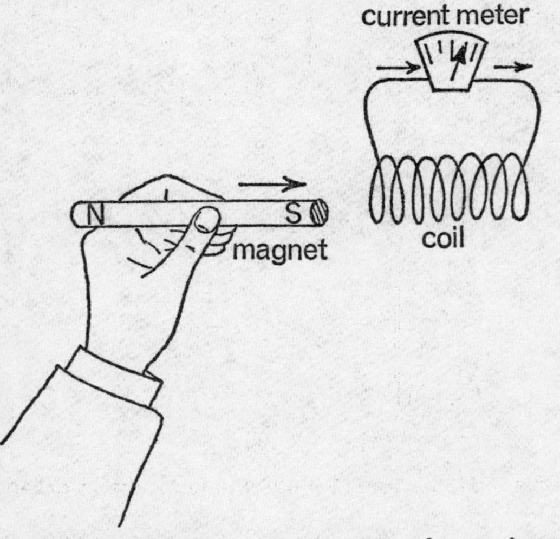

Fig. 60. Inducing current in a coil by means of a moving magnet

which, after seven years of painstaking work, culminated in 1831 in the discovery of electromagnetic induction. We owe to Faraday's discovery the electric generator and the transformer, and consequently our present-day electricity-supply industry.

One of Faraday's basic experiments is very simple and well worth repeating. A coil of wire is wound on an empty cardbard tube (Fig. 60), and the ends of the wire are connected to a galvanometer (or sensitive centre-zero ammeter). If a permanent magnet is plunged into the coil the meter registers a momentary current (the needle kicks to the right, say). When the magnet is pulled out of the coil the meter registers a momentary current in the opposite direction (the needle kicks to the left). No current is registered while the magnet is at rest, either in the coil or outside.

Similar results are obtained if the magnet is fixed and the coil moved towards and away from it. No current is registered unless the coil is in

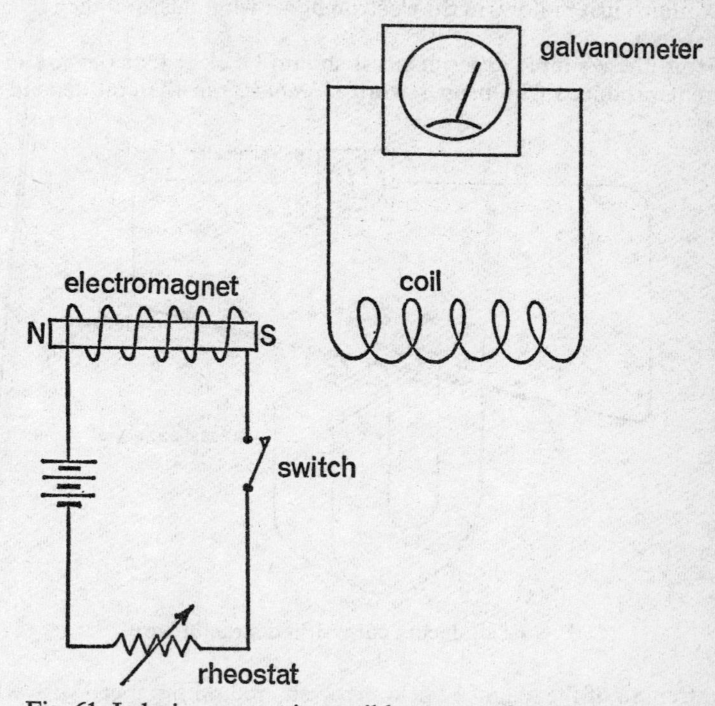

Fig. 61. Inducing current in a coil by means of an electromagnet

motion. It is also apparent that the faster the magnet (or the coil) moves, the greater is the current recorded.

We might be tempted to conclude that motion is an essential component of this experiment, but this is not true, as a further experiment will prove. The permanent magnet is replaced by an electromagnet (Fig. 61) placed near to the fixed coil or, if possible, inside it. If the electromagnet current is increased quickly, by means of the rheostat, the meter again registers a momentary current: if the electromagnet current is decreased quickly the needle kicks in the opposite direction. No current is registered by the meter when the electromagnet current is steady.

Once again the size of the momentary current depends on the speed with which the electromagnet current is varied. In fact, very large 'kicks' are obtained by simply switching on the electromagnet; equally large 'kicks' in the opposite direction occur when the electromagnet is switched off.

The best results of all are produced by adjusting the rheostat so that maximum current flows in the electromagnet windings, and then opening the switch.

From these simple experiments it should be clear that the size of the current produced (the proper word is *induced*) in the coil depends on

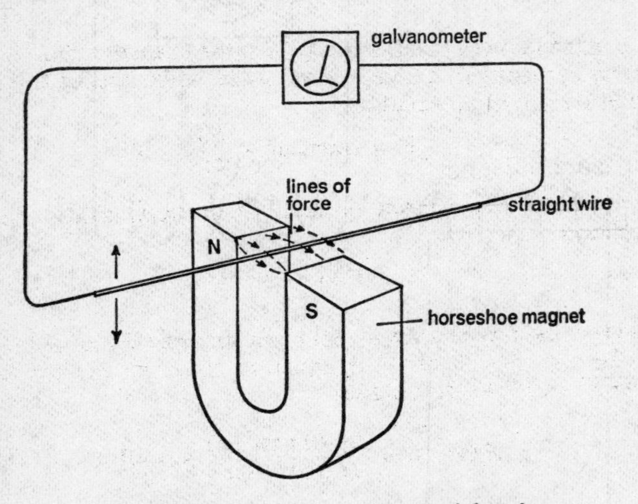

Fig. 62. Inducing current in a straight wire

the strength of the magnetic field involved, and on the speed with which the magnetic field is changed.

We have spoken of the current induced in the coil. Two points need to be clarified here. First, the current exists only when there is a complete circuit, but the e.m.f. which drives the current is present even if the circuit is broken. Thus it is the *induced e.m.f.* that is of primary importance. Secondly, the coil is not essential: a straight wire moved quickly up and down between the poles of a horseshoe magnet has an e.m.f. (and hence a current) induced in it, as shown by the kick of the needle of a galvanometer connected to its ends (Fig. 62).

This latter experiment reveals a further important fact: hardly any e.m.f. is induced when the wire in Fig. 62 is moved quickly from side to side (i.e. along the lines of force). The conductor must cut across the lines of force if an e.m.f. is to be induced in it.

INDUCED E.M.F.

The induced e.m.f. depends on the strength of the magnetic field and on the speed with which the magnetic field is changed. More precisely, we may say that the magnitude of the induced e.m.f. depends on the number of lines of force cut per second. Another way of putting this, and a form which helps in making the statement into an equation, is to say that the magnitude of the induced e.m.f. is proportional to the rate of change of magnetic flux, or

$$E \propto \frac{\Delta\phi}{\Delta t}$$

where E is the induced e.m.f. (volts), $\Delta\phi$ is the change of flux (measured in webers) occurring in a time interval Δt seconds. Putting this into the more useful form of an equation,

$$E = \text{constant} \times \frac{\Delta\phi}{\Delta t}$$

The units have been chosen so that the proportionality constant is unity. Hence

$$E = \frac{\Delta\phi}{\Delta t} \text{ volts} \qquad (1)$$

If, however, the flux change is measured in maxwells, a factor of 10^{-8} has to be introduced into this equation to allow for the fact that 1 maxwell = 10^{-8} webers; hence

$$E \text{ (volts)} = \frac{\Delta\phi \text{ (maxwells)}}{\Delta t \text{ (seconds)}} \times 10^{-8} \qquad (1a)$$

EXAMPLE: A coil composed of 50 turns of wire is cut by 50,000 lines of force (maxwells). If this flux completely disappears (i.e. the field collapses) in 0·01 sec, what is the e.m.f. induced in the coil?

$$\begin{aligned} \text{Initial flux} &= 50{,}000 \text{ maxwells} \\ \text{Final flux} &= 0 \\ \text{Change of flux} &= 50{,}000 \text{ maxwells} \\ &= \Delta\phi \\ \Delta t &= 0{\cdot}01 \text{ sec} \\ E &= \frac{\Delta\phi}{\Delta t} \times 10^{-8} \\ &= \frac{50{,}000}{0{\cdot}01} \times 10^{-8} \\ &= 0{\cdot}05 \text{ V} \end{aligned}$$

This is the e.m.f. induced in one turn of the coil. If all 50 turns are cut by the flux the same e.m.f. (0·05 V) will be induced in each of them. Hence the total e.m.f. between the ends of the coil is

$$50 \times 0{\cdot}05 = 2{\cdot}5 \text{ V} \quad \textit{Answer}$$

EXAMPLE: An insulated wire 10 metres long is stretched between the wingtips of an aeroplane. What is the voltage between its ends when the aircraft is flying horizontally at 500 km per hour in a region where the vertical component of the Earth's magnetic field is 45×10^{-6} webers per square metre?

The wire moves through 500 km every hour. In 1 hour it sweeps out a rectangular area of 500 km $\times$ 10 m $= 5 \times 10^6$ square metres. The flux cutting each square metre is given as 45×10^{-6} webers; hence the flux cut by the wire in 1 hour is $5 \times 10^6 \times 45 \times 10^{-6} = 225$ webers. Using eqn. (1):

$$\Delta\phi = 225 \text{ Wb}$$
$$\Delta t = 3{,}600 \text{ sec}$$
$$E = \frac{\Delta\phi}{\Delta t} \text{ volts}$$
$$= \frac{225}{3{,}600}$$
$$= 0{\cdot}0625 \text{ V} \quad \textit{Answer}$$

LENZ'S LAW

Eqn. (1) gives the magnitude of the e.m.f. induced by a changing magnetic field, but it does not indicate the polarity of the e.m.f. or the direction of the current due to the e.m.f.

The direction of the induced current can be found by experiment. For example, in Fig. 63 the galvanometer needle kicks to the right when the S pole of the magnet is plunged into the coil. A battery (in series with large resistor to protect the meter) can be connected to the galvanometer in place of the coil to show whether a deflection to the right indicates a clockwise current or an anti-clockwise current.

However, there is really no need to carry out such an experiment, since a little common sense will disclose the answer. The end of the coil nearest the magnet becomes *either* a S pole *or* a N pole while the induced current is flowing. If it becomes a N pole it will attract the permanent magnet and suck it into the coil. There would then be no need to plunge the magnet into the coil—the induced current would do this work for us. The greater the attraction, the faster the magnet would move and a greater current would be induced, causing an even greater force of attraction. We could easily devise a perpetual-motion machine based on this experiment. Unfortunately, experience—and the law of conservation of energy—tells us that something is wrong. Our basic assumption that the nearer end of the coil becomes a N pole must be incorrect. In fact, this end becomes a S pole and tends to repel the S pole of the permanent magnet; work has to be done on the magnet to bring it near the coil.

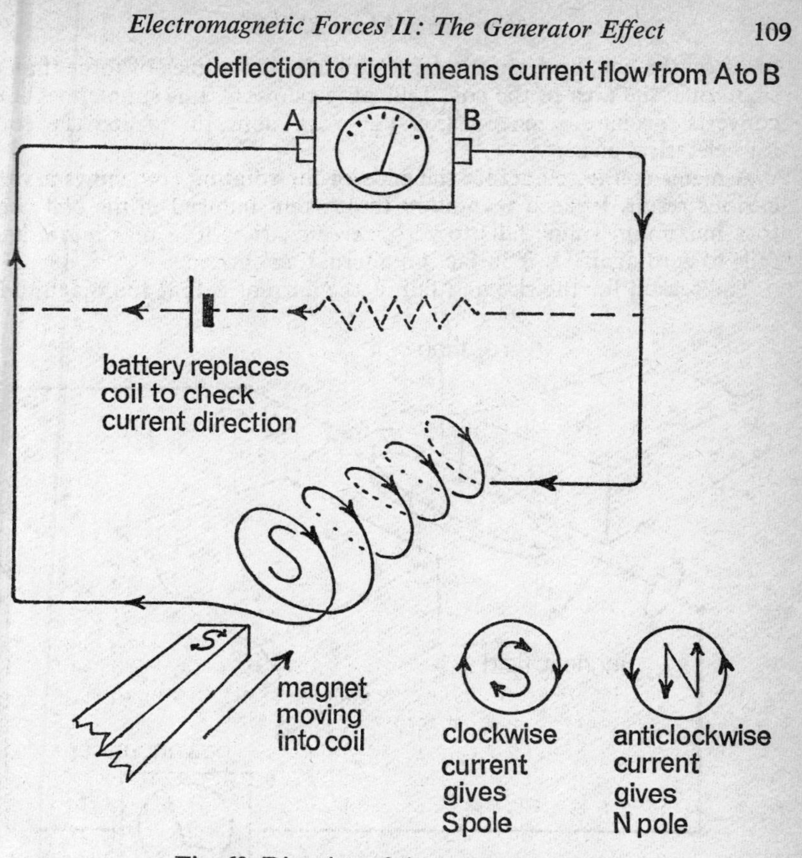

Fig. 63. Direction of the induced current

If the nearer end of the coil is a S pole the induced current must be flowing clockwise around the coil, as shown in Fig. 64.

Such considerations led H. F. Emil Lenz to postulate in 1834 the following law, which bears his name: *an induced current is always in such a direction as to oppose the motion or change causing it.*

THE GENERATOR PRINCIPLE

To induce an e.m.f. or a current requires a magnetic field and a conductor moving at right angles to the direction of the field. The simplest method of achieving this is to arrange for a coil of wire to be rotated continuously in a magnetic field, as shown in Fig. 64. Such an arrangement forms the basis of all types of electromagnetic generator.

The magnitude of the e.m.f. induced in the coil will depend on the strength of the magnetic field, the speed of rotation, the number of turns on the coil, and, since a large coil cuts more lines of force than a small one, the area of the coil. Like all generators, this simple machine converts mechanical energy (i.e. the work done in turning the coil) into electrical energy.

A meter connected across the ends of the rotating coil shows a very curious result. In each revolution the current induced in the coil rises to a maximum value, falls to zero, *reverses*, rises to a maximum, and falls to zero again. It is in fact an alternating current.

The reason for the rise and fall of the current is that the magnitude

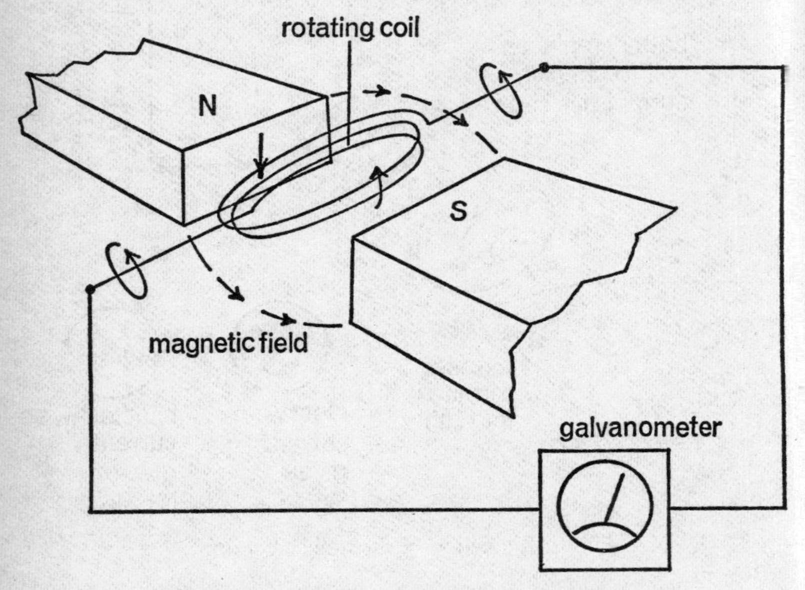

Fig. 64. Generating an e.m.f. in a rotating coil

of the induced e.m.f. depends on the rate at which lines of force are *cut*. When the coil is vertical (Fig. 65) it moves along the lines of force without actually cutting them; no e.m.f. is induced. When the coil is horizontal its movement is at right angles to the field, and the greatest possible number of lines of force is cut per second; the e.m.f. induced reaches its maximum value E_{max}.

The reason for the reversal of the e.m.f. twice in a revolution may be seen in Fig. 64. The side of the coil near the S pole is moving upwards; half a revolution later it has reached the N pole and is moving downwards. Thus twice in every revolution this section of the coil changes direction *relative to the field*, and the same applies to the other half of the coil. As a result, the polarity of the induced e.m.f. is reversed each time the plane of the coil is at right angles to the field (i.e. vertical in Fig. 64), and so is the direction of the induced current.

We could make use of Lenz's law to find the direction of the induced current in each part of the coil, but there is an easier method known as Fleming's *right-hand rule*. (This is very similar to, and easily confused with, the left-hand rule for the force on a conductor given on p. 91.) The right-hand rule states that if the thumb and first two fingers of the right hand are held at right angles to one another, so that the first finger points in the direction of the magnetic field, and the thumb points in

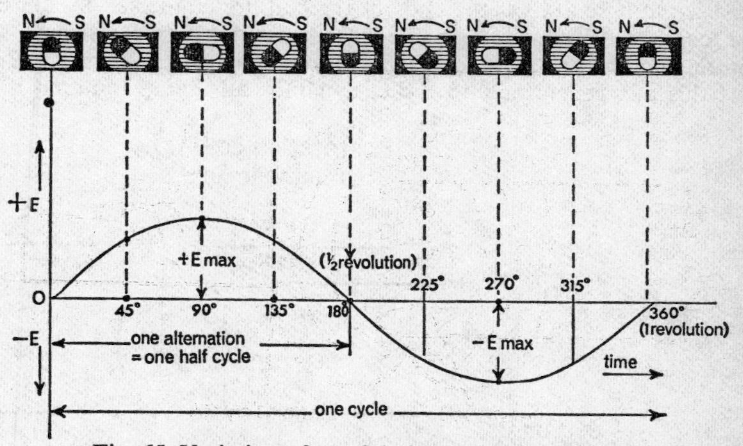

Fig. 65. Variation of e.m.f. induced in a rotating coil

the direction of the motion, then the second finger indicates the direction of the induced current. (It is once again the direction of the conventional positive-to-negative current that is indicated, not the direction of the electron flow.)

Fig. 65 shows the variation of the induced e.m.f. during one complete rotation of the coil; it comprises a positive half-cycle followed by a negative half-cycle. As the coil goes on turning, the pattern is repeated over and over again, each revolution generating one cycle of e.m.f. The number of complete cycles generated in 1 sec is called the *frequency* of the alternating e.m.f., and is, of course, equal to the number of revolutions made by the coil every second.

The waveform of the alternating e.m.f. shown in Fig. 65 is called a *sine wave*, since the value of the e.m.f. corresponding to any position of the coil is given by the formula

$$E = E_{max} \sin \theta$$

where θ is the angle measured between the plane of the coil and the vertical. E_{max} is the maximum value (also called the amplitude) of the e.m.f. occurring when the coil cuts the greatest number of lines of force per second, i.e. when $\theta = 90°$. Sine waves and the equation

$$E = E_{max} \sin \theta$$

will be considered in more detail in connexion with alternating current in Chapters 8–10.

THE D.C. GENERATOR

As we have just seen, a coil rotating in a magnetic field generates an alternating e.m.f.; if the ends of the coil are connected to a circuit the current driven through the circuit is alternating current. The rotating coil is thus a crude a.c. generator. It is a simple matter to convert this a.c. generator into a d.c. dynamo; all that is required is an

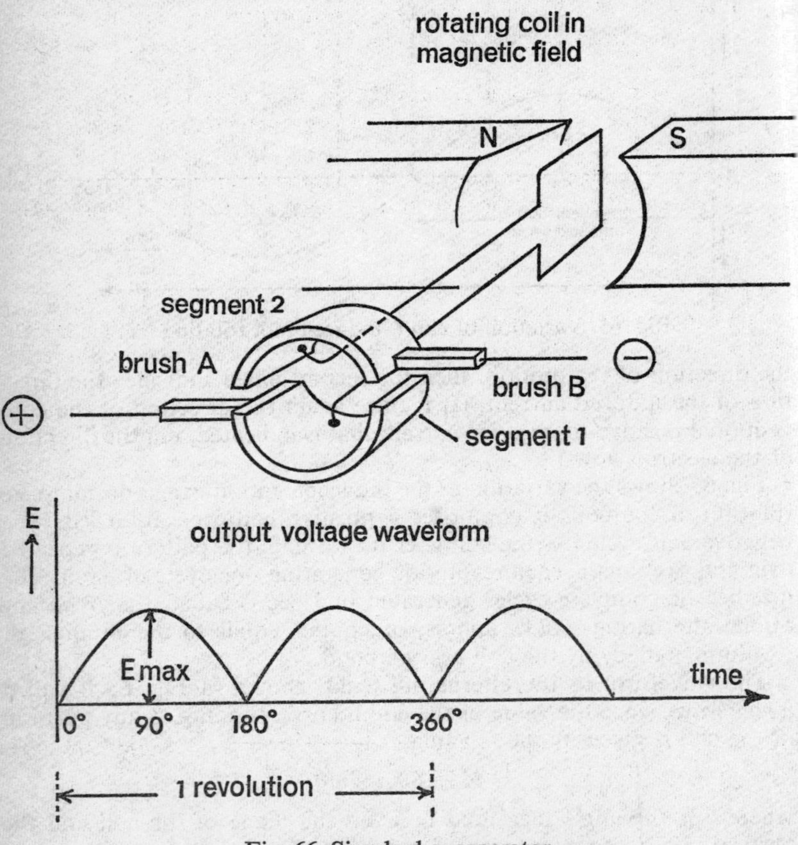

Fig. 66. Simple d.c. generator

automatic switching device to reverse the e.m.f. at the beginning of every *negative* half-cycle.

The switching device is exactly the same as the commutator on a simple d.c. motor (see p. 99). It consists of a ring split into two separate segments, as shown in Fig. 66, each segment being connected to an end of the coil. Stationary blocks of carbon, called brushes, rub on the rotating segments and lead away the induced current.

Just before the coil reaches the vertical position, brush A is in contact with segment 1 and brush B is in contact with segment 2. When the coil reaches the vertical position the direction of the induced e.m.f. reverses. But now brush A is in contact with segment 2 and brush B is in contact with segment 1. So, in spite of the reversal of e.m.f., brush A is still making contact with the positive end of the coil and brush B is still making contact with the negative end. Even though the induced e.m.f. rises and falls, brush A is always positive and brush B is always negative. The output voltage, as shown in Fig. 66, fluctuates but does not alternate.

A much steadier output is obtained by using several coils rotating together with their ends connected to separate segments of the commutator.

If a steady voltage is required the output of the d.c. generator is passed through a smoothing circuit, containing capacitors and choke coils, described on p. 153.

SELF-INDUCTANCE

According to Lenz's law, an induced e.m.f. opposes the change of current that produces it. For this reason, induced e.m.f. is often referred to as 'back e.m.f.'.

Obviously for a given current change some circuits, such as coils, produce a greater back e.m.f. than others, such as straight wires. The characteristic property of a circuit that accounts for the production of an induced voltage or back e.m.f. is its *inductance*. The greater the inductance, the greater is the opposition to current changes, and hence the greater the back e.m.f.

The circuit in Fig. 67 shows a coil of wire connected to a battery. When the switch is suddenly closed, the current rising from zero establishes a magnetic field of increasing intensity, and an increasing number of lines of force cut across the turns of the coil. This induces in the coil a back e.m.f. that opposes the growth of the current and causes it to rise more slowly than it would without the magnetic field. As the field stabilizes, the number of lines of force becomes constant and the back e.m.f. drops to zero. The current then reaches its maximum value

$$I_{\text{max}} = \frac{E}{R}$$

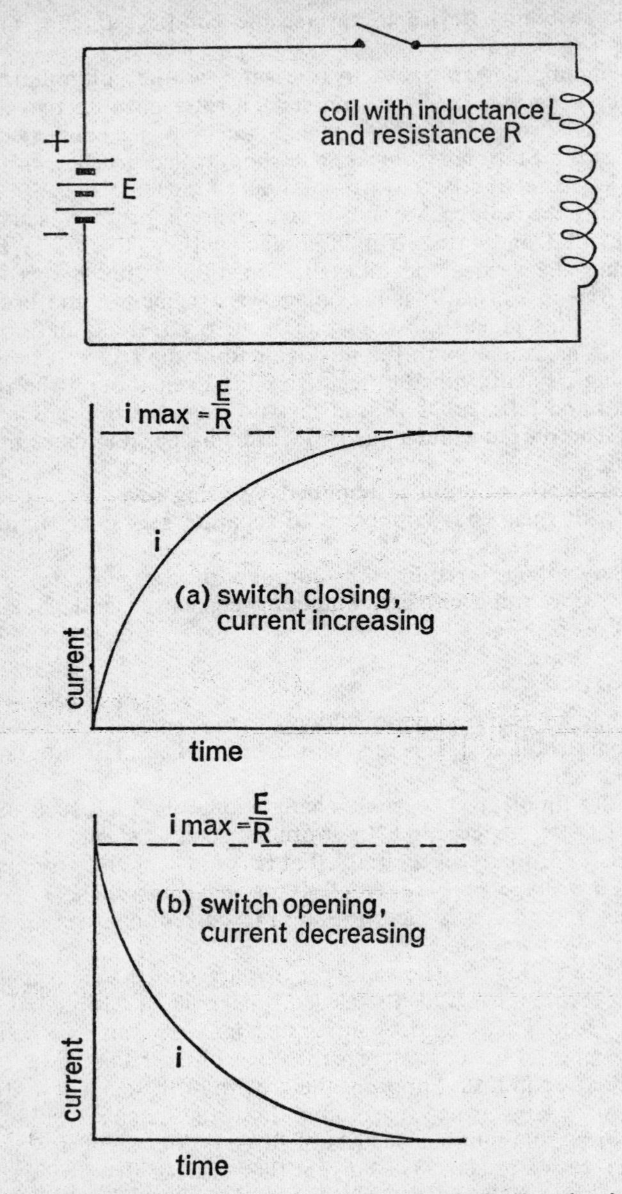

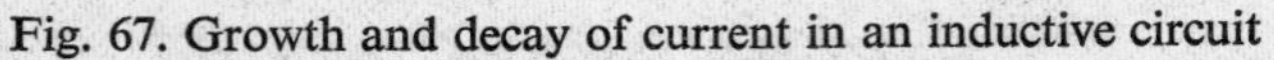
Fig. 67. Growth and decay of current in an inductive circuit

as determined by Ohm's law. The shape of the curve (Fig. 67 (*a*)) and the time the current takes to reach its maximum value are determined by the *self-inductance* of the coil.

If the battery is removed and the ends of the coil connected together

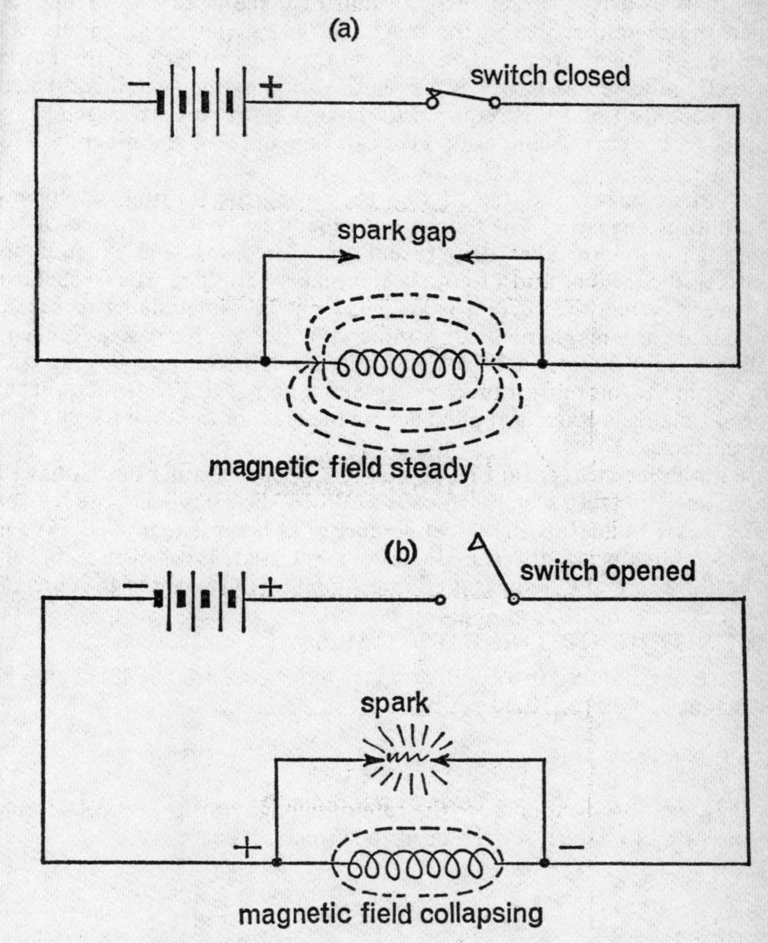

Fig. 68. Energy is stored in an inductive circuit

to form a 'short circuit' the magnetic field will collapse. Again a back e.m.f. is induced, but this time it opposes the decay of the current. The short-circuit current therefore does not immediately fall to zero, but decays in the manner shown in Fig. 67 (*b*). This kind of curve is known as an exponential.

A remarkable demonstration of self-inductance is shown in Fig. 68, where a spark gap is connected across the ends of a coil. When the switch is closed a magnetic field builds up in and around the coil, but there is no spark across the gap because the e.m.f. of the battery is far too small to drive current through that distance in dry air. However, when the switch is opened the magnetic field collapses and induces a back e.m.f. much greater than the e.m.f. of the battery. If the gap is correctly adjusted a spark will be seen as soon as the switch is opened. Since roughly 1,500 V is required to make a spark jump between sharp points 1 cm apart in air, some idea can be gained of the magnitude of the back e.m.f.

A circuit rather like that of Fig. 68 is used for starting the current in a fluorescent lamp. The fluorescent tube is connected in place of the spark gap, a bimetallic strip (usually enclosed inside a neon lamp) replaces the switch, and of course the mains supply is used instead of a battery. When the current is switched on the bimetallic strip breaks the circuit, the magnetic field in the coil (called in this case a 'ballast') collapses and induces a back e.m.f. big enough to ionize the mercury vapour in the fluorescent tube, as explained on p. 46. From then on the normal mains voltage is applied across the ends of the tube to maintain the current.

It should be clear from Fig. 68 that a coil, or any other circuit having inductance, acts as a *store of energy*. When the circuit is switched on the current builds up slowly while energy is being 'stockpiled'. When the switch is opened the stored energy is released: most of it reappears as heat in the spark, some is used in prolonging the decay of the current.

MAGNITUDE OF THE BACK E.M.F.

The e.m.f. induced in a conductor is proportional to the rate of change of magnetic flux (p. 107) or

$$E \propto \frac{\Delta\phi}{\Delta t}$$

In the case of self-inductance the rate of change of flux is proportional to the rate of change of current in the circuit; hence

$$E \propto \frac{\Delta I}{\Delta t}$$

where ΔI (amps) is the small change in current occurring in the small time interval Δt (sec). Turning this into an equation:

$$E = \text{constant} \times \frac{\Delta I}{\Delta t}$$

$$= L\frac{\Delta I}{\Delta t}$$

The proportionality constant L is called the coefficient of self-induction, or simply the inductance. To show that the back e.m.f. opposes the applied voltage, it is usual to put a minus sign in this equation, thus

$$E = -L\frac{\Delta I}{\Delta t} \qquad (2)$$

The unit of inductance is the *henry*, named after Joseph Henry, an American who discovered independently the laws of electromagnetic induction at almost the same time as Faraday. If a current change of 1 amp per second induces an e.m.f. of 1 V in a circuit, the self-inductance of that circuit is 1 henry (1 H).

EXAMPLE: What is the self-inductance of a coil in which an e.m.f. of 20 V is induced by a current change of 200 mA in 0·1 sec?

$$E = 20\text{ V}$$
$$\Delta I = 200\text{ mA} = 0{\cdot}2\text{ A}$$
$$\Delta t = 0{\cdot}1\text{ sec}$$
$$E = -L\frac{\Delta I}{\Delta t}$$

disregarding the minus sign,

$$20 = L \times \frac{0{\cdot}2}{0{\cdot}1}$$
$$L = \tfrac{20}{2}$$
$$= 10\text{ henrys} \quad \textit{Answer}$$

MUTUAL INDUCTANCE

When the current changes in an inductive circuit, e.g. a coil, the resulting change of flux induces a back e.m.f. not only in the circuit itself but also in any other circuit that happens to be cut by the flux.

Fig. 69 (*a*) shows the build up of flux around a coil when current is switched on. This induces a back e.m.f. which opposes the current. When the switch is opened the magnetic field collapses and the decreasing flux induces another back e.m.f. The field extends beyond the coil, however, and if a second coil is cut by the flux while it is growing or decaying an e.m.f. will be induced in the second coil as well as in the first. A galvanometer connected to the second coil (Fig. 69 (*b*)) records a momentary current when the switch connected to the primary coil is opened or closed—even though the two coils are quite separate.

This inductance between separate circuits is known as *mutual* inductance. The size of the e.m.f. E induced in the secondary is given by the formula

$$E = -M\frac{\Delta I}{\Delta t} \qquad (3)$$

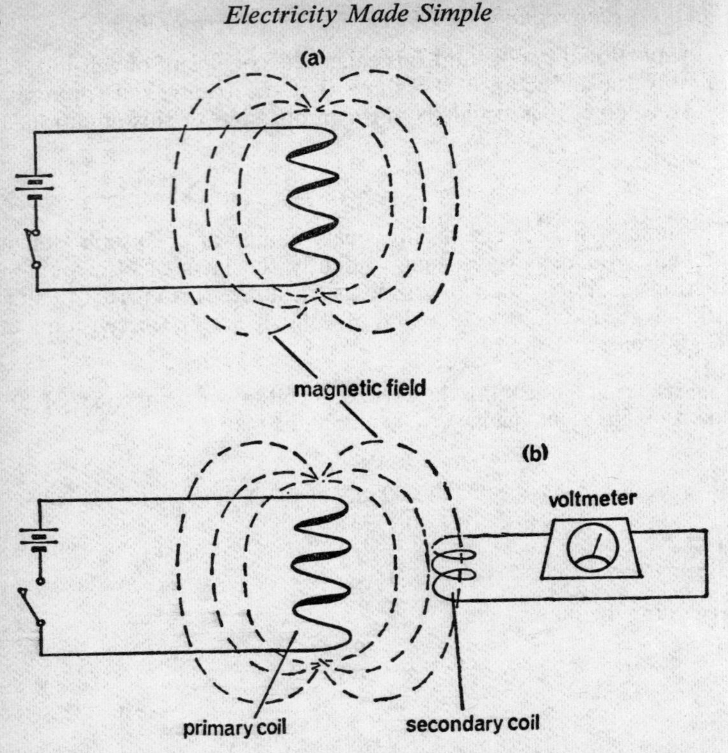

Fig. 69. Mutual inductance: changing flux in one coil induces an e.m.f. in the other

where M is the coefficient of mutual inductance (measured in henrys), ΔI is the change in current in the *primary* circuit occurring in a time Δt.

EXAMPLE: An e.m.f. of 30 V is induced in one of a pair of coils when the current in the other changes at a rate of 5 amps per second. What is the coefficient of mutual inductance?

$$E = 30 \text{ V}$$
$$\Delta I = 5 \text{ A}$$
$$\Delta t = 1 \text{ sec}$$
$$E = -M\frac{\Delta I}{\Delta t}$$

disregarding the minus sign,

$$30 = M \times 5/1$$
$$M = 6 \text{ henrys} \quad \textit{Answer}$$

If a current change of 1 amp per second in one coil induces an e.m.f. of 1 V in another coil the two coils have a mutual inductance of 1 henry. The mutual inductance of two circuits depends on their individual self-inductances and on the degree of coupling between them. Mathematically, mutual inductance is expressed by the relation

$$M = k\sqrt{L_1 L_2}$$

where L_1 and L_2 are the self-inductances of the primary and secondary circuits, respectively, and k is the 'coefficient of coupling'.

The value of k lies between 0 and 1. If all the flux produced by the primary circuit cuts through the secondary circuit, $k = 1$. This is the tightest possible coupling between the circuits. At the other extreme, if none of the flux cuts the secondary circuit, $k = 0$.

EXAMPLE: Two coils whose self-inductances are 3 henrys and 12 henrys, respectively, have a mutual inductance of 4 henrys. What is their coefficient of coupling?

$$M = 4 \text{ H}$$
$$L_1 = 3 \text{ H}$$
$$L_2 = 12 \text{ H}$$
$$M = k\sqrt{L_1 L_1}$$
$$4 = k\sqrt{3 \times 12}$$
$$= k\sqrt{36}$$
$$= k \times 6$$
$$k = \tfrac{4}{6} = 0{\cdot}666 \text{ (no units)} \quad \textit{Answer}$$

The most important application of the mutual-inductance principle is found in the transformer used for stepping up and stepping down *alternating* voltages. Transformers are essentially a.c. devices—in fact, they are very liable to be damaged if connected to a d.c. supply—and will be discussed in Chapter Eight on alternating current. At this stage it is sufficient to describe a transformer as a primary coil and a secondary coil, electrically isolated from each other but having a high coefficient of coupling and mutual inductance.

A kind of transformer that does work from a d.c. source is the spark coil used in motor-car ignition systems. It consists of an iron core on which is wound a primary coil of a few turns of thick wire. A secondary coil of several thousand turns of thin wire is wound over the primary. The direct current supplied by the car battery is fed to the primary via a mechanical interrupting device called a contact-breaker. Hence the current in the primary is a rapidly fluctuating current, and the flux in the iron core grows and decays equally rapidly. The e.m.f. induced in the secondary, about 20,000 V, is applied to the sparking plug in order to ignite the petrol mixture. In a car travelling at 60 m.p.h. the coil must furnish about 200 sparks every second.

EDDY CURRENTS

Any conductor moving through a magnetic field will have an e.m.f. induced in it, and the induced e.m.f. will cause currents to circulate if the conductor is large enough. These circulating currents are called *eddy currents*.

Eddy currents are frequently set up in the rotating cores of electrical machines such as motors and generators; they also occur in the cores of transformers, caused not by motion but by the changing flux. They are objectionable, since they tend to heat up the iron in which they circulate, and in any case are a waste of energy.

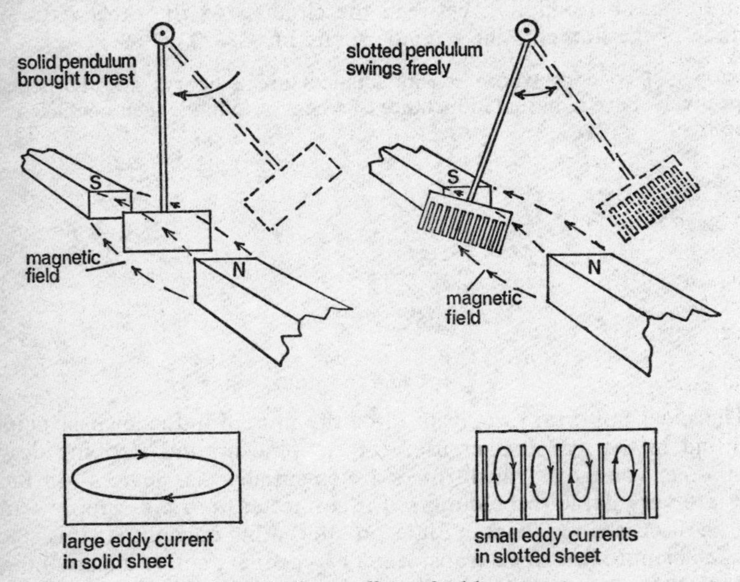

Fig. 70. Braking effect of eddy currents

By Lenz's law, eddy currents in a conductor oppose its motion through a magnetic field and thus exert a braking effect. A pendulum consisting of an aluminium sheet (Fig. 70) hung in a magnetic field will not swing freely because of the interaction between the applied magnetic field and the magnetic field of the eddy currents. But the pendulum will swing freely if slots are cut in it, as shown. Although the same e.m.f. is induced as before, there is now no easy path for currents to circulate, and the braking effect is much reduced.

Eddy-current braking is utilized in the domestic kilowatt-hour meter shown in Fig. 23. The permanent magnet mounted astride the rotating disc exerts a resisting torque that is proportional to the speed of rotation. With no resisting torque the disc would tend to accelerate, in which case its speed would not be proportional to the product of current × voltage (see p. 44).

Dividing the iron core of a transformer or electrical machine into thin sheets (called laminations) has the same effect as cutting slots in the pendulum mentioned above. The magnetic properties of the core are practically unaffected, but the electrical resistance (perpendicular to the plane of the laminations) is very greatly increased. If the laminations are insulated from one another with a thin coating of varnish eddy currents can almost be eliminated.

SUMMARY

A conductor in a *changing* magnetic field has an e.m.f. induced in it. The magnitude of the induced e.m.f. is proportional to the rate of change of magnetic flux: $E = \frac{\Delta\phi}{\Delta t}$ volts, where $\Delta\phi$ is the flux change in webers occurring in a time Δt sec. In a closed circuit the induced e.m.f. causes a current whose direction is such as to oppose the motion or change producing the e.m.f. (Lenz's law).

A coil of wire rotated in a magnetic field has an alternating e.m.f. induced in it; the size of the e.m.f. at any instant is given by $E = E_{max} \sin \theta$, where θ is the angle between the plane of the coil and a line perpendicular to the field.

A form of direct current is obtained from a rotating coil fitted with a commutator to reverse the e.m.f. at the beginning of each negative half-cycle.

The e.m.f. induced by a variation of current in any circuit is

$$E = -L\frac{\Delta I}{\Delta t}$$

where ΔI is the change in current occurring in a time Δt and L is the coefficient of *self-inductance*; L is measured in henrys.

A changing current in one circuit will induce an e.m.f. in any nearby circuit, even though the two are not connected electrically. The magnitude of the induced e.m.f. in the secondary circuit is given by

$$E = -M\frac{\Delta I}{\Delta t}$$

where M is the coefficient of *mutual inductance* and $\frac{\Delta I}{\Delta t}$ is the rate of change of current in the primary circuit.

$$M = k\sqrt{L_1 L_2}$$

where L_1 is the self-inductance of the primary circuit, L_2 is the self-inductance of the secondary circuit, and k is the coefficient of coupling between the two circuits.

Eddy currents are small circulating currents set up in a large conductor either by the motion of the conductor in a magnetic field or by a varying magnetic field in a stationary conductor. They are reduced by splitting the conductor into laminations.

CHAPTER EIGHT

ALTERNATING CURRENT

An alternating current is one in which the electron flow changes direction periodically. So far we have dealt only with direct-current (d.c.) circuits where the electrons flow steadily in the same direction all the time. Because of their tremendous technical importance, we must now turn our attention to alternating-current (a.c.) circuits, in which the electrons oscillate backwards and forwards.

We have already met (p. 111) the concept of an alternating e.m.f. developed when a coil of wire is rotated in a magnetic field. This e.m.f. rises from zero to a maximum value E_{max}, then falls to zero again before increasing to a maximum with reversed polarity. The magnitude of the rising, falling, and reversing e.m.f. varies with the position of the coil according to the equation

$$E = E_{max} \sin \theta$$

The voltage of the a.c. mains supply varies in precisely the same way.

In Great Britain and most of Europe the frequency of the a.c. mains is 50 Hz (cycles per second)*, and the mains voltage goes through a complete cycle (0, E_{max} 0, $-E_{max}$ 0) 50 times every second. The current which this alternating voltage drives through a circuit connected to the mains varies at the same rate of 50 complete cycles (0, I_{max} 0, $-I_{max}$ 0) every second. The magnitude of this alternating current at any instant is given by

$$I = I_{max} \sin \theta$$

It might be asked why alternating current is almost universally used in preference to direct current for the mains supply. The answer is very largely concerned with the ease with which the voltage of an alternating supply can be changed. This, as we shall see, is of prime importance in transmitting power efficiently from the generating station to the consumer. There are also practical difficulties associated with generating very large quantities of d.c. power: for example, we cannot at present generate 750 megawatts of power in the form of direct current, but it is quite possible to produce an equivalent amount of power using alternating-current techniques.

For most practical purposes, alternating current is no better and no worse than direct current. It cannot be used for electrolysis or electro-

* The term cycle per second (c/s) has largely been replaced by the term hertz, symbol Hz; the two are identical in every way.

plating, but is very suitable for running electric heaters and electric motors.

THE TRANSFORMER

The inductance of a d.c. circuit (p. 113) is of interest only when the current is varying, and in practice current variations are unlikely to occur except briefly when the current is switched on or off. With an alternating current, however, the magnitude of the current varies all the time. Hence when a.c. is supplied to a coil a back e.m.f. is induced *all the time*. The induced e.m.f. is, of course, an alternating e.m.f., and its frequency is the same as that of the applied alternating current.

The magnitude of the induced e.m.f. depends on the (alternating) voltage applied to the coil and on the coefficient of inductance. This is the principle of the transformer.

The usual type of transformer consists of two coils coupled by mutual inductance (see p. 117). The coils are electrically insulated from each other, but are wound on the same core: they are therefore linked by a common magnetic flux.

A current in one of the coils (the primary winding) produces in the iron core a magnetic flux, almost all of which passes through the other coil (the secondary winding). The core (Fig. 71) is built up from thin strips of special silicon steel (which behaves magnetically like soft iron) insulated from one another by layers of varnish. This laminated construction is designed to reduce eddy currents.

An alternating current in the primary creates an alternating magnetic flux in the core, which in turn induces an alternating e.m.f. in the secondary. If there is no circuit connected to the secondary the ratio of the induced e.m.f. E_s in the secondary to the voltage E_p applied to the primary is the same as the ratio of the number of turns in the two windings, i.e.

$$\frac{E_s}{E_p} = \frac{N_s}{N_p} \qquad (1)$$

where N_s is the number of secondary turns and N_p is the number of primary turns.

Eqn. (1) assumes that every turn of the secondary is linked by the magnetic flux created in the primary (i.e. no flux leakage). It obviously does not apply to air-cored high-frequency transformers used in radio and television.

In a step-up transformer there are more turns in the secondary than in the primary; E_s is greater than E_p, and the output voltage is greater than the input.

In a step-down transformer (Fig. 71 (*a*)) there are fewer turns in the secondary than the primary; E_s is less than E_p, and the output voltage is less than the input.

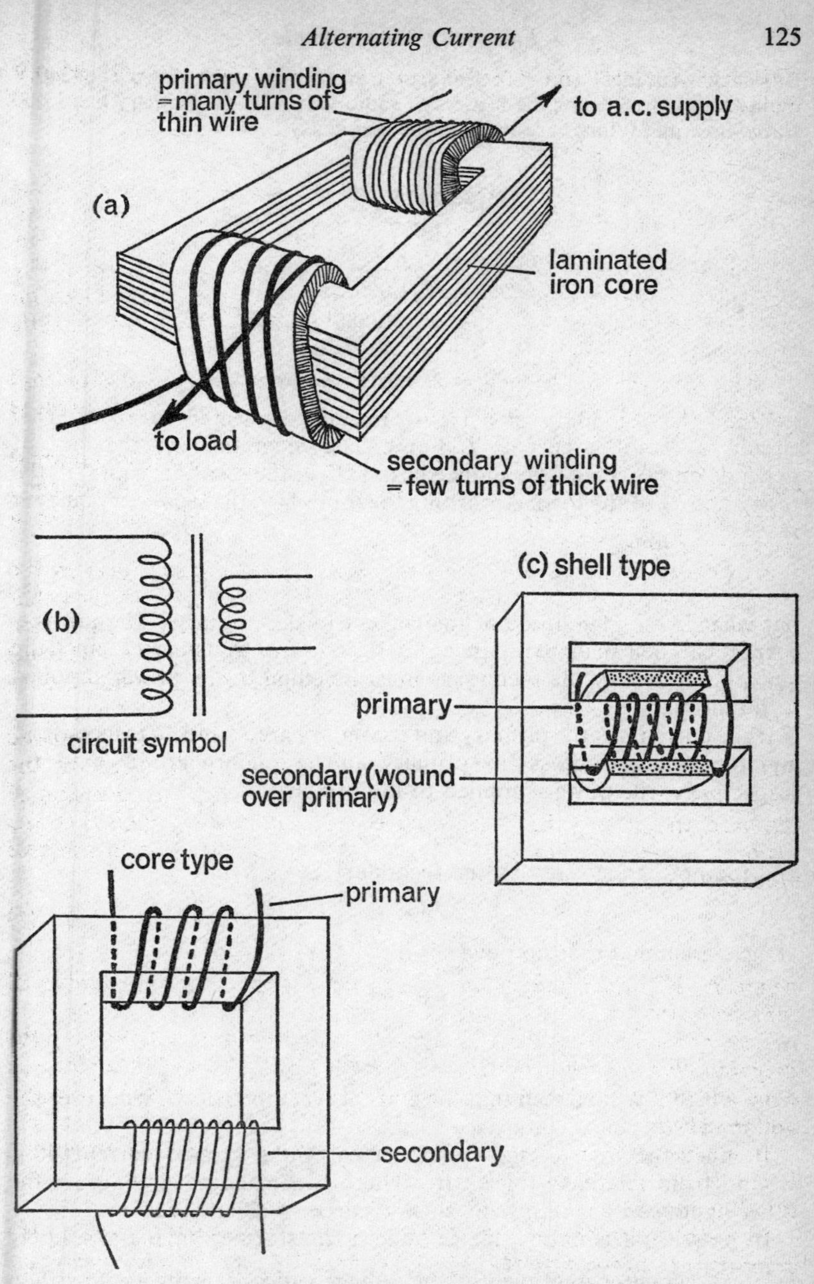

Fig. 71. Construction of an electrical transformer

EXAMPLE: An ideal (no flux leakage) transformer steps down the 240 V mains supply to 6 V for the heaters of radio valves. If the primary has 1,000 turns, how many turns are there in the secondary?

$$E_s = 6\text{ V}$$
$$E_p = 240\text{ V}$$
$$N_p = 1{,}000$$
$$E_s/E_p = N_s/N_p$$
$$6/240 = N_s/1{,}000$$
$$N_s = \frac{6 \times 1{,}000}{240}$$
$$= 25\text{ turns} \quad \textit{Answer}$$

When the secondary of a transformer is connected to an external circuit or 'load' a current I_s flows. The magnitude of the current depends on the size of the induced e.m.f. E_s in the secondary and on the resistance R_L of the load. According to Ohm's law, the secondary current is

$$I_s = \frac{E_s}{R_L}$$

but what is the magnitude of the current in the primary? The primary current can be calculated quite easily for an *ideal* transformer, since the *power* delivered by the secondary must be equal to the power supplied to the primary

If the currents in the primary and secondary are I_p and I_s, respectively, and the voltages across the primary and secondary are E_p and E_s, respectively, the power supplied to the primary is

$$I_p \times E_p$$

and the power delivered by the secondary is

$$I_s \times E_s$$

Hence, assuming that no power is lost,

$$I_p \times E_p = I_s \times E_s$$

or

$$\frac{I_p}{I_s} = \frac{E_s}{E_p} \qquad (2$$

from which it will be seen that the current is stepped down whenever the voltage is stepped up, and vice versa.

Because the low-voltage coil must carry the greater current, it is wound from relatively thick wire. The higher-voltage winding, on the other hand, can use thin wire, since it carries a smaller current.

In practice, it is impossible to build a transformer having absolutely no flux leakage. Consequently, the voltage ratio $\frac{E_s}{E_p}$ is always less than the turns ratio $\frac{N_s}{N_p}$. Actual transformer windings, moreover, have some resistance. Hence there is a voltage drop ($I \times R$) across the coil: the voltage across the primary is a little less than the applied voltage, and the voltage available from the secondary is a little less than the induced e.m.f. A certain amount of power is turned into heat and wasted in overcoming the resistance of the windings; this is called I^2R loss (sometimes called copper loss—except, of course, where the windings are made of aluminium).

Power is also lost in the transformer core. Some is wasted in setting up eddy currents (p. 120) and some, called the hysteresis loss, is wasted aligning the magnetic domains in the core material. The sum of the eddy-current loss and the hysteresis loss is called the core loss (sometimes called iron loss).

All these losses add up to prevent the output power equalling the power supplied to the primary. The actual ratio of the output power to the input power gives the *efficiency* of a transformer. Hence eqn. (2) has to be modified for practical transformers to

$$\frac{E_s}{E_p} = \frac{I_p}{I_s} \times \text{Efficiency} \qquad (3)$$

EXAMPLE: What is the efficiency of a transformer which delivers a current of 9 A at 24 V when its primary draws a current of 2 A from a 120-V supply?

$$E_s = 24 \text{ V}$$
$$E_p = 120 \text{ V}$$
$$I_p = 2 \text{ A}$$
$$I_s = 9 \text{ A}$$
$$\frac{E_s}{E_p} = \frac{I_p}{I_s} \times \text{Efficiency}$$
$$\frac{24}{120} = \frac{2}{9} \times \text{Efficiency}$$
$$\text{Efficiency} = \frac{24}{120} \times \frac{9}{2}$$
$$= \frac{9}{10}$$
$$= 90 \text{ per cent} \quad \textit{Answer}$$

Actual transformers can realize efficiencies of 90–98 per cent, the higher figure applying to large commercial types.

Apart from their use in changing voltages, transformers are of value in isolating one circuit from another. For example, a power tool, such as a drill, may be run from an isolating transformer having a 1 : 1 turns ratio. There is then no direct connexion between the tool and the mains, and the risk of electric shock is lessened considerably. A 10:1 turns ratio would provide even greater protection.

Unlike the more usual type of transformer described above, the *auto-*

transformer (Fig. 72) has its primary and secondary combined into a single winding. In the step-down version the whole coil forms the primary and is connected to the a.c. supply. A section of the coil is 'tapped off' to form the secondary, and the reduced output voltage is obtained from the ends of this section.

If the a.c. supply is connected to the ends of the tapped-off section of

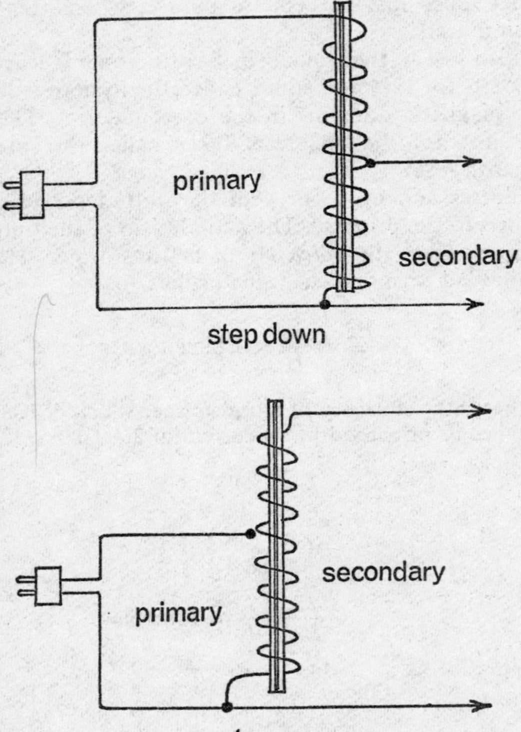

Fig. 72. Schematic of the autotransformer

the coil a stepped-up voltage is obtained from the ends of the whole coil.

The big advantage of the autotransformer is that, by moving the position of the 'tapping point', the turns ratio can be varied to any desired value. The coil is wound on a laminated iron ring, but the windings, although insulated from the core, are of bare wire. A sliding contact sweeps over the windings in the same manner as the sliding contact of a wire-wound variable resistor.

POWER TRANSMISSION

The generators in a power station produce an e.m.f. of up to 20,000 V (20 kV). If, say, 20 megawatts (20×10^6 W) are to be sent out to consumers at this voltage the current would be 1,000 amps $\left(\text{Current} = \frac{\text{Power}}{\text{Voltage}}\right)$.

Since the heating effect of current is proportional to I^2R (see p. 40), the cables through which the 1,000-amp current is distributed must be of very low resistance (i.e. of enormous cross-sectional area), or an intolerable amount of energy will be turned into heat and wasted. But if the voltage is stepped up by means of an efficient transformer to, say,

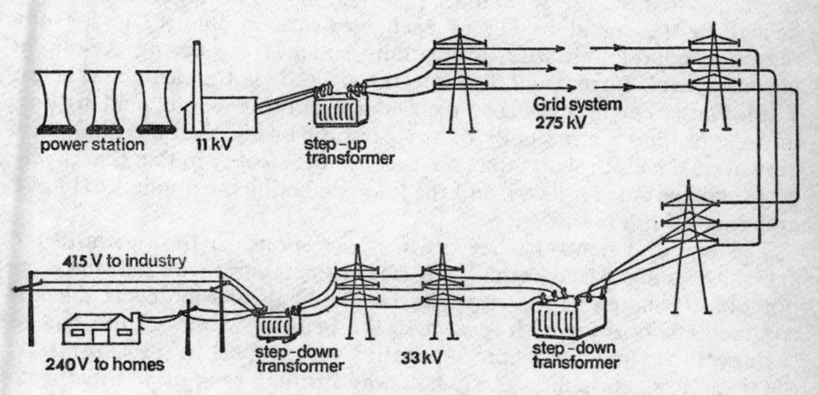

Fig. 73. High-voltage distribution of power

200 kV a current of 100 A would distribute the same 20 megawatts of power, and the heating effect for a given size of cable is reduced by a factor of $\frac{(1{,}000)^2}{(100)^2} = 100$.

This is why power is distributed through the wires of the national Grid system at very high voltages (up to 400 kV), as shown in Fig. 73. In this typical example the generator e.m.f. of 11 kV is stepped up to 275 kV for long-distance transmission from the power station, which may be sited on a coalfield to reduce transport costs, to a city hundreds of miles away. At the consumers' end of the Grid the voltage is stepped down to 33 kV at a substation and finally to 240 V at a local transformer for domestic use.

The transformers used in the Grid handle very large currents and have to be kept cool by circulating dry oil around the windings. The core is built up of silicon-steel laminations insulated with kaolin or varnish. The conductors themselves are covered with paper and in some types of transformer are varnished after winding. The whole structure is thoroughly dried in a vacuum before the cooling oil is poured in.

In practice, power is distributed through the Grid as 3-phase current (see below) and has to be converted into the more familiar single-phase form of alternating current for domestic use.

Despite the fact that the cable adopted for house-wiring contains three conductors, it carries a single alternating current. The three wires are LINE, NEUTRAL, and EARTH. In Great Britain the insulation covering the wires is coded brown for line, blue for neutral, and green/yellow for earth.

The earth wire carries no current. It is connected at one end to a plate buried in the ground (it should not be connected to a water pipe, since so many of these are nowadays non-metallic). The other end is connected to the metal casing of each electrical appliance. If for any reason a conducting wire accidentally touches the casing so that it becomes 'live', current will flow into the ground via the earth conductor. Without the earth conductor the leakage current would tend to flow into the ground via the body of any person who touched the appliance. However, the earth wire offers an easier path: so easy in fact that a very large leakage current flows, and the fuse protecting the appliance 'blows' and cuts off the mains supply.

The line and neutral wires do not correspond to the outward and return wires in a battery circuit. Normally the neutral wire is at the same potential (voltage) as the earth: in the first half of each cycle the line conductor is positive with respect to the neutral; in the second half of each cycle the line conductor is negative with respect to the neutral. At the beginning, end, and exactly half-way through each cycle there is no difference of potential between the line and neutral wires. For a mains frequency of 50 Hz this no-voltage condition occurs every $\frac{1}{100}$ sec (in the United States and Canada, where the mains frequency is 60 Hz, the voltage is, of course, zero every $\frac{1}{120}$ sec).

The fact that the current is also zero twice in every cycle means that there is less risk of sparking between the contacts of a switch, compared with a direct current of equal voltage. Switches and fuses, it should be noted, are inserted into the line (brown) conductor.

THREE-PHASE CURRENT

As its name suggests, a 3-phase supply comprises three separate alternating e.m.f.s whose waveforms are out of step with one another (Fig. 74). Such an e.m.f. is generated when three coils arranged symmetrically about a common axis are rotated in a magnetic field. Industrial consumers are usually supplied with 3-phase power, because it is particularly useful for driving large electric motors (see p. 189).

It might be thought that three separate alternating voltages would call for three pairs of conductors or six wires in all. In fact, three of the six wires, one per phase, can be connected together so that only four wires are required: the fourth wire is the neutral. The four-wire method of 3-phase distribution is shown in the lower part of Fig. 75. The 'loads' to be supplied (usually the windings of a motor) shown as resistors in the diagram are connected between the neutral and each of the other three wires. In this 'star' connexion, as it is called, the potential

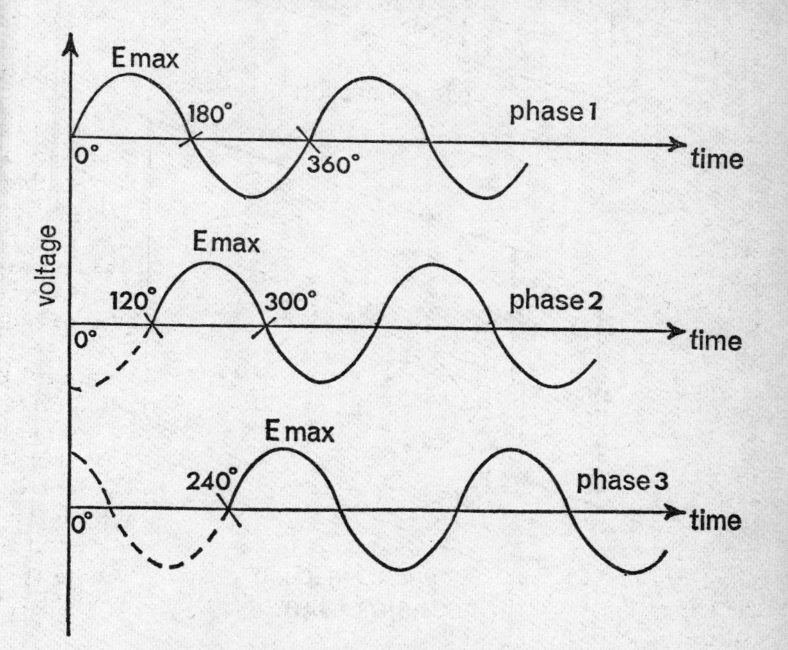

Fig. 74. Voltage waveforms of a 3-phase supply

difference across each load is the supply voltage (typically 415 V) divided by $\sqrt{3}$.

The neutral conductor carries no current and may be dispensed with, leaving the three-wire system shown in the upper part of Fig. 75. The 'loads', shown as resistors, are inserted between each pair of wires, in what is known as a 'delta' connexion.

SKIN EFFECT

A circuit carrying an alternating current is subject to inductance effects which do not arise with direct currents. One example is the back e.m.f. induced even in a straight wire, which, as Lenz's law states, tends

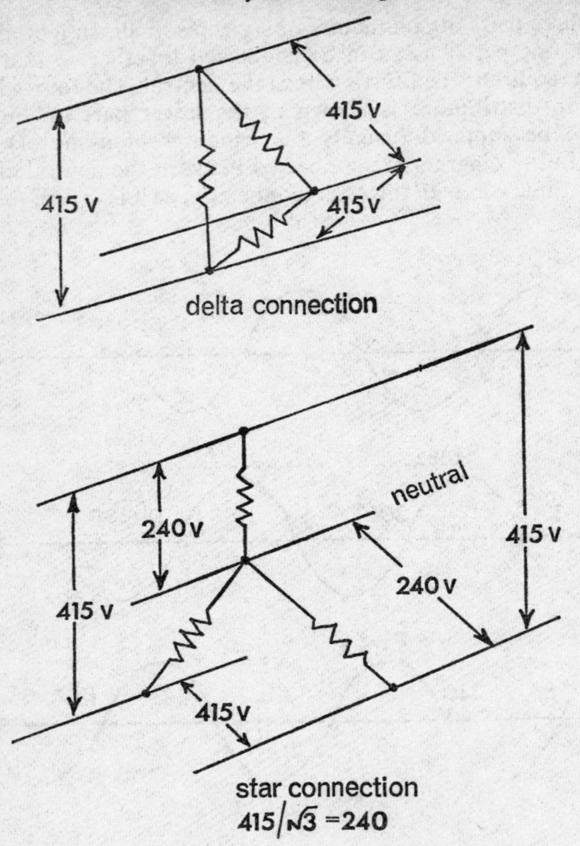

Fig. 75. Delta and star connexions to a 3-phase supply

to oppose the current producing it. The curious aspect of the self-induced e.m.f. is that it is greater along the centre of a conductor than near the surface. Consequently, opposition to the current is greatest near the centre of the wire, and the majority of the current tends to flow close to the surface, where the opposition is least. This phenomenon is known as the 'skin effect'.

Because of the skin effect, the central portion of a conductor carrying a high-frequency current is almost redundant and could be removed without greatly increasing the resistance. The greater the diameter, the more pronounced is the effect. This has led to the use of conductors made of a bundle of fine wires twisted together so as to run alternately on the outside and along the centre of the bundle.

RESISTANCE IN A.C. CIRCUITS

Except at very high frequencies, such as are encountered in radio engineering, where the skin effect becomes appreciable, a resistor behaves in exactly the same way with alternating current as it does with direct current. This assumes, of course, that it is pure, i.e. has no inductance. The equations and formulae discussed in connexion with d.c. circuits apply equally to resistors in a.c. circuits.

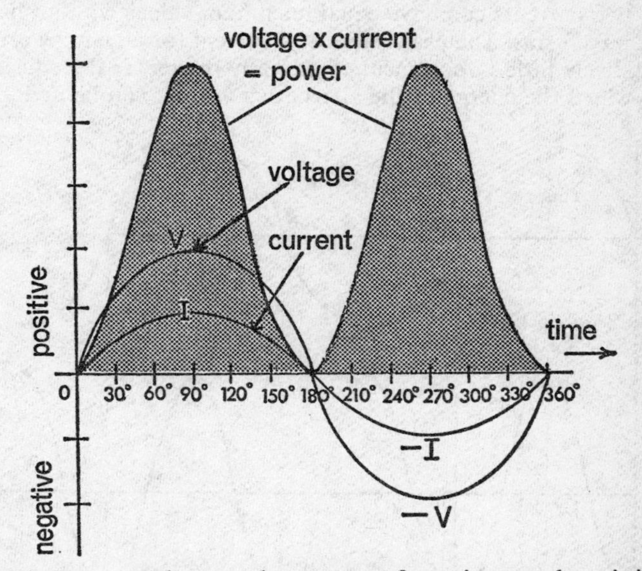

Fig. 76. Current, voltage, and power waveforms in a purely resistive a.c. circuit

The combined resistance of a number of resistors in series is the sum of the individual resistances:

$$R = R_1 + R_2 + R_3 + \quad \text{etc.}$$

The combined resistance R of a number of resistors in parallel is given by

$$\frac{1}{R} = \frac{1}{R_1} + \frac{1}{R_2} + \frac{1}{R_3} + \quad \text{etc.}$$

The power converted into heat when an alternating current flows through a purely resistive circuit is the product of the current and voltage, i.e. $(I_{max} \sin \theta) \times (E_{max} \sin \theta)$ as shown in Fig. 76. In the second half of each cycle the voltage is negative and the current is negative: their product $(-I \times -V)$ is therefore positive, just as it is in the first half of each cycle. The power in a purely resistive a.c. circuit therefore varies between zero and a maximum value twice in every cycle, but it is always positive.

R.M.S. VALUES

An alternating current whose peak value (I_{max}) is, say, 5 A obviously does not produce the same heating effect as a direct current of 5 A: for most of each cycle the alternating current is much less than 5 A. Moreover, two currents can have equal maximum values without having the same waveforms. The peak value of a current (or an e.m.f.) is therefore of relatively little importance; what really matters is the *effective* value from which the energy of the alternating current can be deduced.

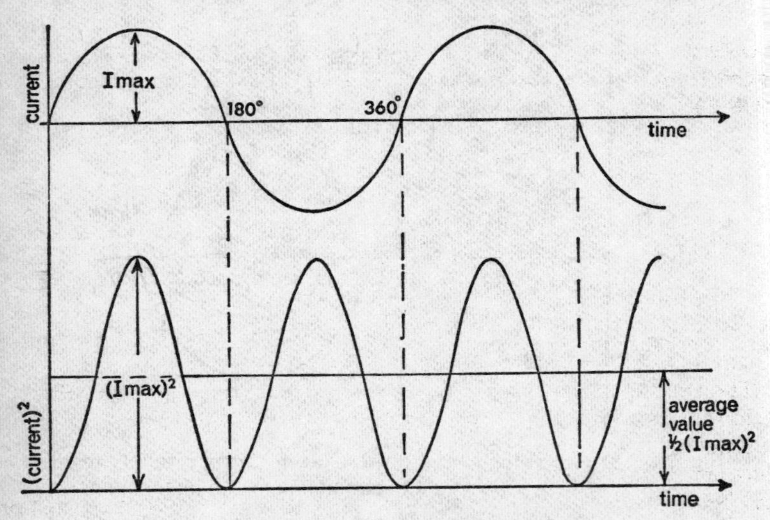

Fig. 77. Current and (current)² waveforms

Accordingly, the *effective* value of an alternating current is defined as that value which produces the same heating effect as a direct current of equal magnitude. In other words, an effective alternating current of, say, 5 A produces the same heating effect as a direct current of 5 A.

The heat produced by a current is proportional to the *square* of the current, as explained in p. 40. In the case of an alternating current, the heat produced is therefore proportional to $(I_{max}\sin\theta)^2$. A graph of $(I_{max}\sin\theta)^2$ is plotted in Fig. 77; it varies between 0 and $(I_{max})^2$ twice per cycle, and the heating effect varies in the same way. The average or mean value of the heating effect is therefore proportional to $\frac{0 + (I_{max})^2}{2}$ or $\frac{1}{2}(I_{max})^2$.

The heating effect of the equivalent direct current is proportional to I^2. Hence, if the two currents produce the same heating effect,

$$I^2 = \tfrac{1}{2}(I_{max})^2$$
$$I = \sqrt{\tfrac{1}{2}(I_{max})^2}$$
$$= \frac{I_{max}}{\sqrt{2}}$$

or the *effective* current is the maximum current divided by $\sqrt{2}$. Similarly, the *effective* e.m.f. is the maximum e.m.f. divided by $\sqrt{2}$, or $E = \frac{E_{max}}{\sqrt{2}}$.

Effective values of current and voltage are usually called 'root-mean-square' values or r.m.s. values.

Unless otherwise stated, any figures quoted for alternating currents and voltages are r.m.s. values. Thus when we refer to the 240-V mains we mean that the r.m.s. value of the voltage is 240 V: the maximum or peak value is

$$E_{max} = E \times \sqrt{2}$$
$$= 240 \times 1{\cdot}414$$
$$= 339 \text{ V}$$

(This calculation is based on the assumption that the mains voltage follows a sine curve; it does not apply to rectified a.c.)

HOT-WIRE AMMETER

A current-measuring instrument known as the hot-wire ammeter is shown in Fig. 78. Because its action depends on the heating effect, it can be used for measuring both alternating current and direct current.

The current to be measured passes through the thin resistance wire AB, which becomes heated and expands. As the wire expands, the slack is taken up by a thread held taut by a spring and passing over a pulley coupled to the needle of the ammeter.

The hot-wire ammeter is not very accurate, but it is useful for measuring very high (radio) frequency alternating currents. The more common 50-Hz currents are usually measured by first converting them to direct current with the aid of a *rectifier* and then using a moving-coil ammeter (p. 95).

RECTIFICATION OF A.C.

A rectifier is the name given to any device which will convert to-and-fro alternating current into one-way direct current, a process known as *rectification* of alternating current. An ideal rectifier may be thought of as a switch which closes a circuit during the positive half of each cycle, and opens to break the circuit during the negative half of each cycle.

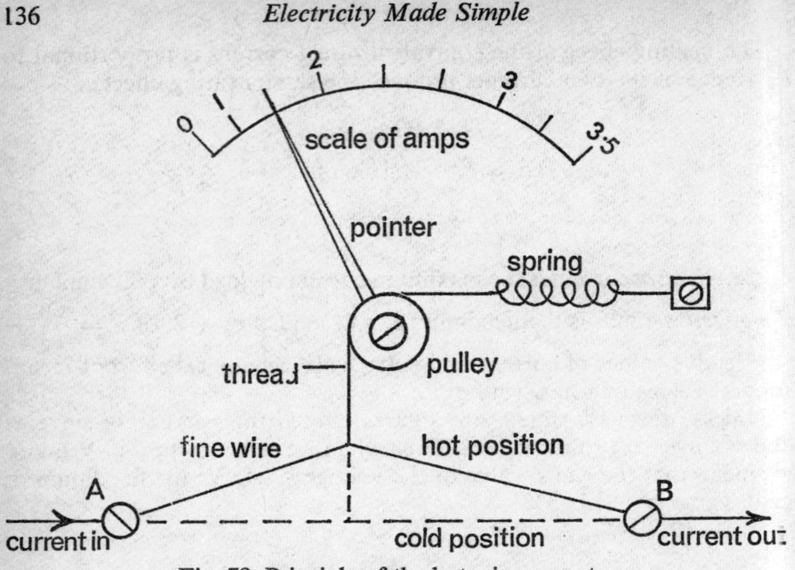

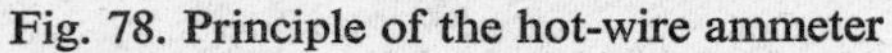
Fig. 78. Principle of the hot-wire ammeter

A diode, either a thermionic valve or a semiconductor, fulfils this switching requirement admirably.

The thermionic diode consists of two metal conductors in a vacuum. One of the conductors, called the cathode, is heated by a thin filament carrying a current; the other conductor, called the anode or plate, is not heated. The free electrons in the heated cathode acquire so much energy that they leave the surface of the metal and form a cloud just outside it. If an alternating voltage is applied between anode and cathode the anode becomes positive with respect to the cathode in the first half of each cycle: the cloud of electrons is then attracted across the valve to the anode (Fig. 79 (*a*)). In the second half of each cycle the anode becomes negative with respect to the cathode, and therefore repels electrons back to the cathode: there is then no current through the valve (Fig. 79 (*b*)).

Where heavy currents are to be rectified, the mercury-vapour valve is preferable to the vacuum diode. A small amount of liquid mercury contained in the valve vaporizes as the cathode is heated. When a positive half-cycle of alternating current is applied to the anode, electrons are attracted from the cathode. In passing across the valve these electrons collide with mercury atoms, and each collision produces more free electrons. The effect is cumulative, so that the current through the valve is several times as great as that of an equivalent high-vacuum diode.

A semiconductor diode requires no heater and has the additional advantage that its resistance (in the forward direction) is generally less than that of a valve. Silicon and germanium are the semiconductors most used for rectifiers.

A germanium diode is essentially a *p–n* junction, comprising a crystal of *n*-type germanium (i.e. germanium that has been 'doped' with a trace

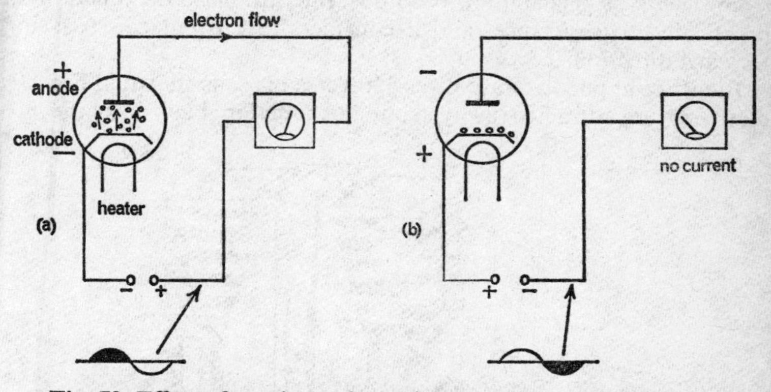

Fig. 79. Effect of an alternating voltage applied to a diode valve

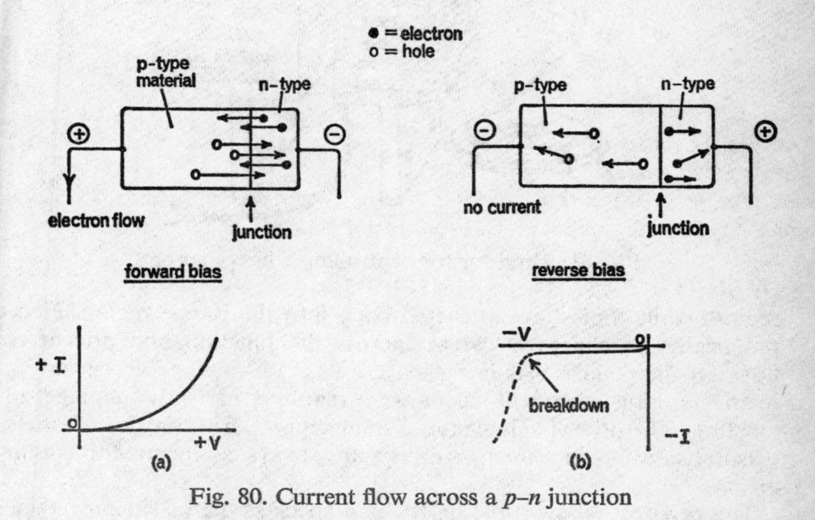

Fig. 80. Current flow across a *p–n* junction

of antimony or arsenic) containing a piece of *p*-type germanium (i.e. germanium that has been doped with a trace of indium or gallium). *N*-type germanium has an excess of free electrons, and *p*-type germanium has a deficit of free electrons (described as an excess of *holes*).

It might be expected that surplus electrons from the *n*-type germanium would immediately flow into the *p*-type region to fill the holes there; but this does not happen, since there is a 'potential barrier' of perhaps 0·1 V at the junction. If a voltage exceeding 0·1 V is applied to the diode, as shown in Fig. 80 (*a*), electrons are attracted across the junction from the *n*-type germanium, while 'holes' are attracted across the junction from the *p*-type germanium. With this 'forward bias', the junction offers practically no resistance, and a considerable current can pass in the forward direction.

But if the applied voltage gives a 'reverse bias', as shown in Fig. 80 (*b*), electrons are attracted away from the junction back into the *n*-type

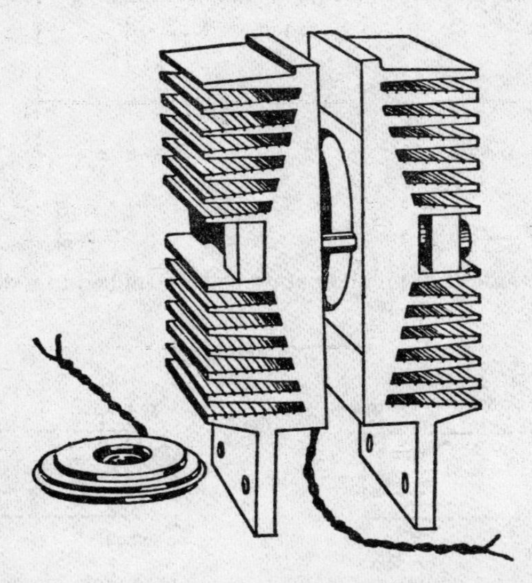

Fig. 81. Thyristor for controlling a heavy current

region, while 'holes' are attracted back into the *p*-type region. Hence practically no charge is carried across the junction and no current flows. If the reverse bias is excessive a sudden 'breakdown current' of relatively large size will occur and may permanently damage the junction. The other likely cause of damage is overheating, and for this reason heavy-current junction diodes are always equipped with cooling fins.

The *thyristor*, shown in Fig. 81, is a special type of silicon rectifier consisting of a *p–n* junction incorporating a separate contact known as a 'gate'. The addition of the gate drastically changes the characteristics of the junction. When a thyristor is operated with reversed bias it blocks the flow of current (unless the voltage becomes excessive and a breakdown current occurs); but unlike conventional rectifiers, a thyristor also blocks the flow of current when operated with a forward bias. Only when a certain critical value of the (forward) voltage is reached does the device switch suddenly to a highly conducting state.

In practice, the thyristor is operated with a forward bias somewhat less than the critical (breakover) voltage, and is 'turned on' by a small pulse of current applied to the gate, via the twisted leads shown in Fig. 81.

HALF-WAVE RECTIFIER

A diode of any kind inserted in an a.c. circuit, as shown in Fig. 82, presents a very high resistance to current in one direction and a low resistance to current in the other. Hence only the positive half-cycles

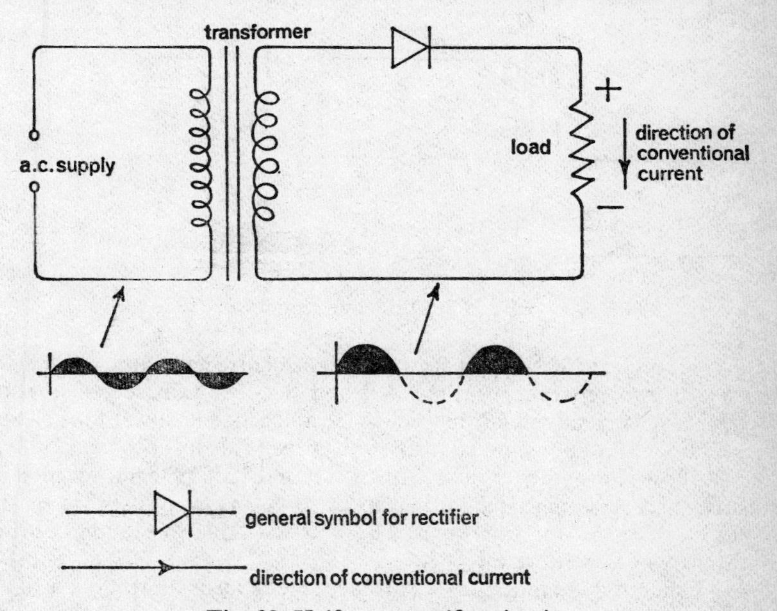

Fig. 82. Half-wave rectifier circuit

of voltage will be able to make a current flow through the diode; in effect the device 'rectifies' by filtering out the negative half-cycles. The 'rectified' current passing through the load is not steady, but at least it flows in only one direction. Such a current would be satisfactory for charging storage batteries, but it would cause an intolerable 'mains hum' if used for running a radio receiver. For the latter purpose, current variations can be ironed out by means of a smoothing circuit (see p. 153).

The easiest way of looking at a rectifier circuit diagram, such as Fig. 82, is to regard the triangle-and-bar symbol as an arrow indicating the path of conventional (positive-to-negative) current, i.e. to imagine that positive half-cycles alone can pass from triangle to bar.

The simple diode circuit illustrated in Fig. 82 is a half-wave rectifier, since only half of the a.c. waveform is usefully employed.

FULL-WAVE RECTIFIER

A more efficient circuit with two diodes turns both halves of each a.c. cycle into a form of direct current. This is known as a full-wave rectifier (Fig. 83).

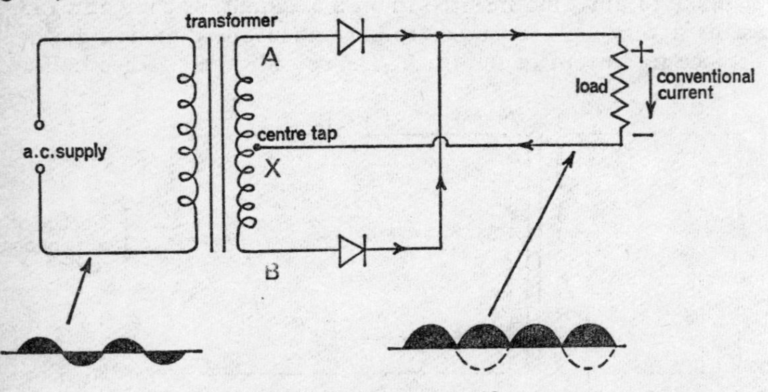

Fig. 83. Full-wave rectifier

The secondary of the transformer has a centre tap, and can be regarded as two separate windings AX and BX. In the first half of each a.c. cycle A is positive with respect to X, and current flows through the upper diode. At the same time B is negative with respect to X, so no current flows in the lower diode. In the second half of each cycle B is positive and A is negative with respect to X, so current flows in the lower diode but not in the upper. Thus current flows in the load during both halves of each cycle.

An incidental advantage of this arrangement is that no rectified current passes through the transformer windings: a direct current might saturate the core and make it less effective by lowering its permeability.

BRIDGE CIRCUIT

The expense of a centre-tapped transformer is avoided by the four-diode bridge circuit shown in Fig. 84, which is another full-wave rectifier.

During the first half of each cycle the top of the transformer secondary winding is positive with respect to the bottom: point A is therefore positive, and current flows through diode 4. At the same time the bottom of the secondary is negative with respect to the top: point B is therefore negative, and current flows through diode 2 in the direction CB.

During the second half of each cycle the top of the secondary is negative with respect to the bottom: point A is therefore negative, and current flows through diode 1 in the direction CA. At the same time the bottom of the secondary is positive with respect to the top: point B is therefore positive and current flows through diode 3.

Hence current flows in two out of the four diodes during every half cycle. The output voltage, derived from the full voltage of the trans-

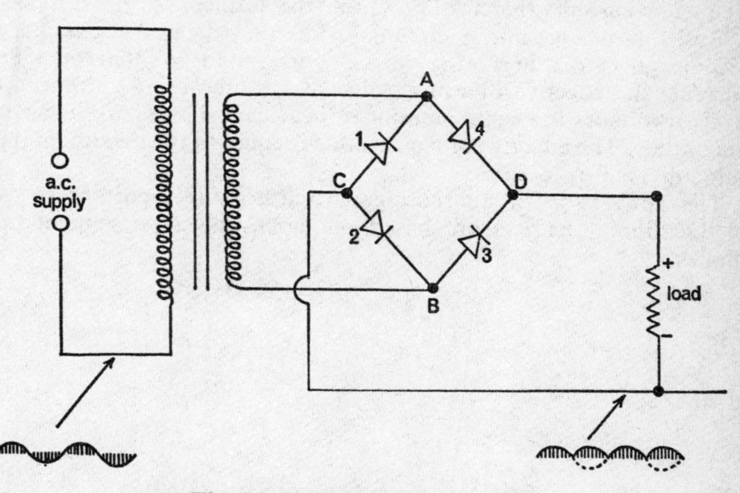

Fig. 84. Full-wave bridge rectifier

former secondary, is theoretically twice that of the circuit in Fig. 83, where the transformer secondary is effectively cut in two by the centre tap.

SUMMARY

The voltage supplied by the a.c. mains at any given instant is $E = E_{max} \sin \theta$, and the current that this voltage produces in a resistance is $I = I_{max} \sin \theta$.

The frequency is the number of complete cycles of current or voltage occurring in 1 sec. Each complete cycle consists of a positive half-cycle followed by a negative half-cycle.

Alternating voltages can be stepped up and stepped down by means of a transformer. The ratio of the secondary voltage to the primary voltage is equal to the ratio of the secondary turns to the primary turns.

The ratio of the primary current to the secondary current is theoretically equal to the ratio of the secondary voltage to the primary voltage. In a normal transformer the primary winding is electrically separate from the secondary; in an autotransformer the primary and secondary are combined in a single winding.

To reduce heat losses (proportional to I^2R), power is transmitted at high voltages. A 3-phase supply has three voltage waveforms displaced from one another at an angle of 120°.

At very high frequencies the centre of a conductor offers relatively more opposition to current than does the rest of the conductor; high-frequency currents therefore flow near the surface.

Resistors behave in a.c. circuits just as they do in d.c. circuits. The heating effect of alternating current corresponds to that of a direct current: the effective or r.m.s. value of an alternating current is that which produces the same amount of heat as a direct current of equal magnitude. The effective or r.m.s. value is equal to the maximum (peak) value divided by $\sqrt{2}$.

The conversion of alternating current to direct current is called rectification. Rectifiers are based on diode valves or semiconductor diodes.

CHAPTER NINE

CAPACITORS AND INDUCTORS

When currents are steady, the effects of capacitors and inductors are not apparent: no changes are going on. In a.c. circuits currents rise and fall all the time, and the effects of capacitors and inductors become very pronounced—so much so that an entire chapter has to be devoted to them.

CAPACITORS

A capacitor is essentially a device for storing charge. In practice, it is made from two conductors, called plates, separated by an insulator called the dielectric. If a battery is connected to the capacitor shown in Fig. 85 electrons flow from the negative terminal on to the right-hand plate, which thus acquires a negative charge. Meanwhile electrons are attracted from the left-hand plate to the battery; the left-hand plate therefore acquires a deficiency of electrons, which is described as a positive charge. The electron flow in each connecting wire stops as soon as the plates have been charged to the same voltage as the battery. No current passes between the plates, of course, because they are separated by the dielectric, in this case air. Charges are stored on the plates even if the battery is removed, until a conducting path is provided by a circuit joining one plate to the other.

How much charge is stored depends on the potential difference between the plates, and this is equal to the terminal voltage of the battery. If the stored charge is Q coulombs (i.e. $+Q$ on one plate and $-Q$ on the other) and the p.d. between the plates is V volts we can write

$$Q \propto V$$

or

$$Q = CV$$

where the proportionality constant C is called the *capacitance* of the system.

Capacitance is measured in *farads*. One farad (1 F) is the capacitance of a system in which one coulomb of charge is stored when the p.d. is one volt. But the farad turns out to be too large a unit for most purposes, so it is divided into a million microfarads, and a billion picofarads (picofarads, symbol pF, are sometimes called micro-microfarads, symbol μμF).

$$1\ \text{F} = 10^{6}\ \mu\text{F} = 10^{12}\ \text{pF}$$

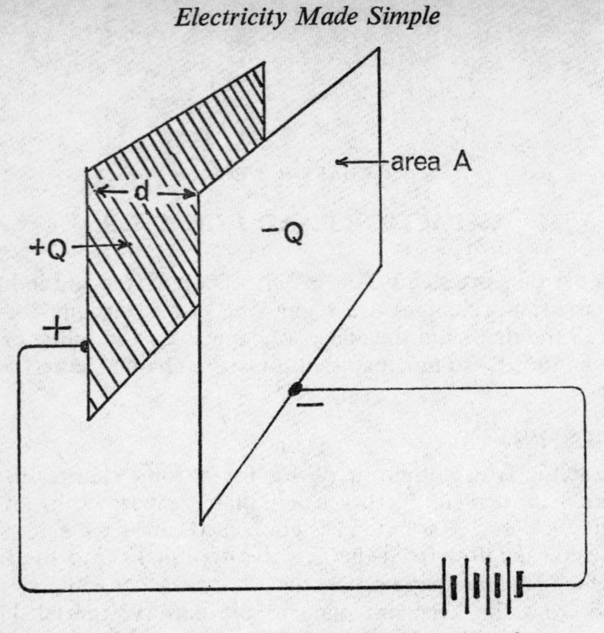

Fig. 85. Principle of the parallel-plate capacitor

The magnitude of the capacitance is governed by the area of the plates, by their distance apart, and by the nature of the dielectric. Using symbols,

$$C = \text{constant} \times \frac{\varepsilon_r A}{d}$$

where A is the overlapping area common to both plates, d is the distance between the plates, and ε_r is the 'relative permittivity' of the dielectric (also called the 'dielectric constant'). The constant of the proportion-

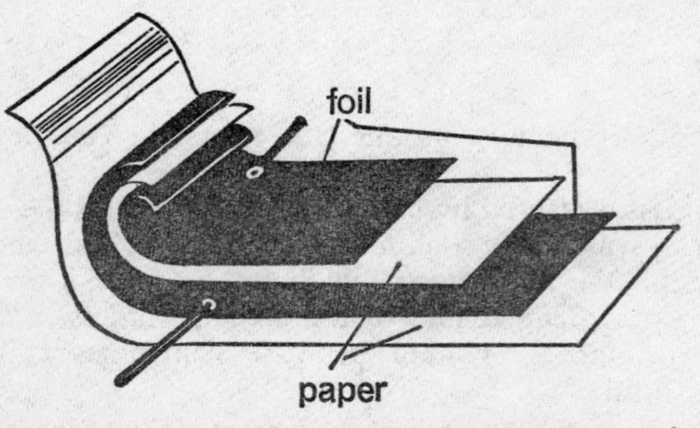

Fig. 86. Construction of a paper-dielectric capacitor

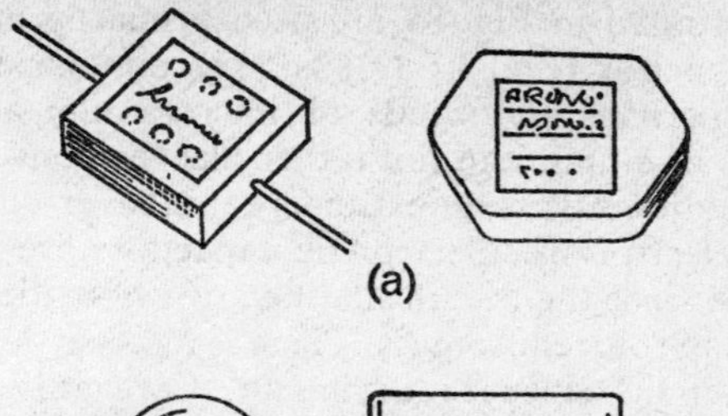

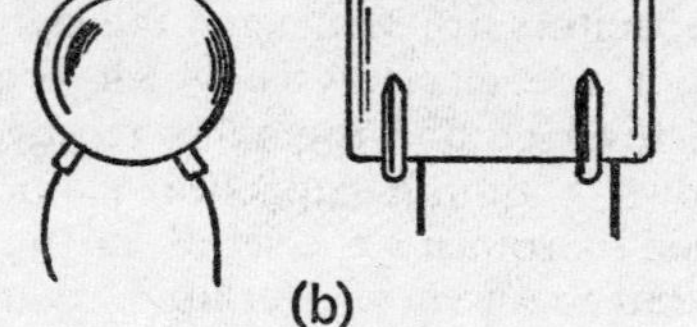

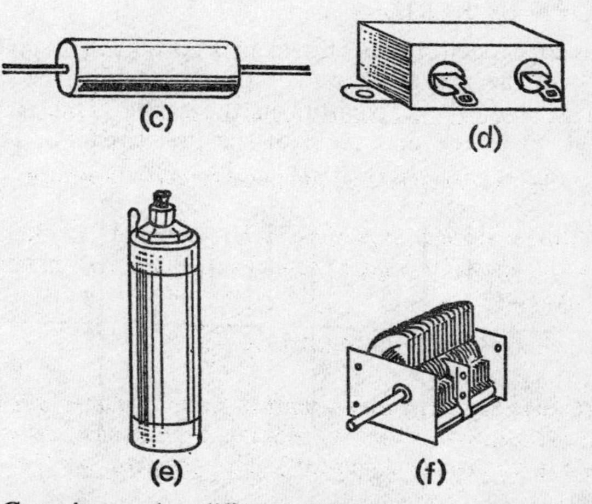

Fig. 87. Capacitors using different dielectrics: (*a*) mica; (*b*) ceramic; (*c*) plastic film; (*d*) waxed paper; (*e*) aluminium oxide; (*f*) air

ality in the equation depends on the shape of the plates and on the units in which A and d are measured.

Since $C \propto \frac{\varepsilon_r A}{d}$, large values of capacitance are obtained by using plates of large area held close together, and by choosing a dielectric of high relative permittivity such as polystyrene ($\varepsilon_r = 2{\cdot}6$), glass ($\varepsilon_r = 4$–$7$), mica ($\varepsilon_r = 5$–$9$), or waxed paper ($\varepsilon_r = 3$), compared with air ($\varepsilon_r = 1$). The practicable capacitor shown in Fig. 86 uses plates made of aluminium foil and a dielectric of waxed paper. By rolling the 'sandwich' tightly, a large capacitance is achieved with minimum bulk. The various capacitors illustrated in Fig. 87 are used primarily in electronics. The electrolytic capacitor (Fig. 87 (*e*)) is very popular where extra-large capacitances (perhaps thousands of microfarads) are required. The plates are rolls of aluminium foil rather like the construction described above, but the dielectric is an extremely thin layer of aluminium oxide generated by electrolysis. Electrolytic capacitors are not as reliable as the other types, since the thin insulating layer of oxide may break down and allow leakage current to pass. There is also a risk of the electrolyte, which is sealed in the capacitor, drying out. Care must be taken to ensure that electrolytic capacitors are connected the right way round (i.e. with correct polarity), or the electrolysis which originally generated the oxide layer will be reversed and the oxide removed.

Fig. 87 (*f*) shows a tuning capacitor from a radio receiver. Its capacitance is varied by rotating the set of movable plates out of the set of fixed plates so that the area of overlap is reduced to zero.

CAPACITORS IN SERIES

If a series of capacitors, as shown in Fig. 88 (*a*), is given a charge Q (i.e. $-Q$ on the first plate and $+Q$ on the last) the same charge appears on each of the capacitors in the series. This is because the electrons put on to the first plate of the first capacitor (C_1) repel an

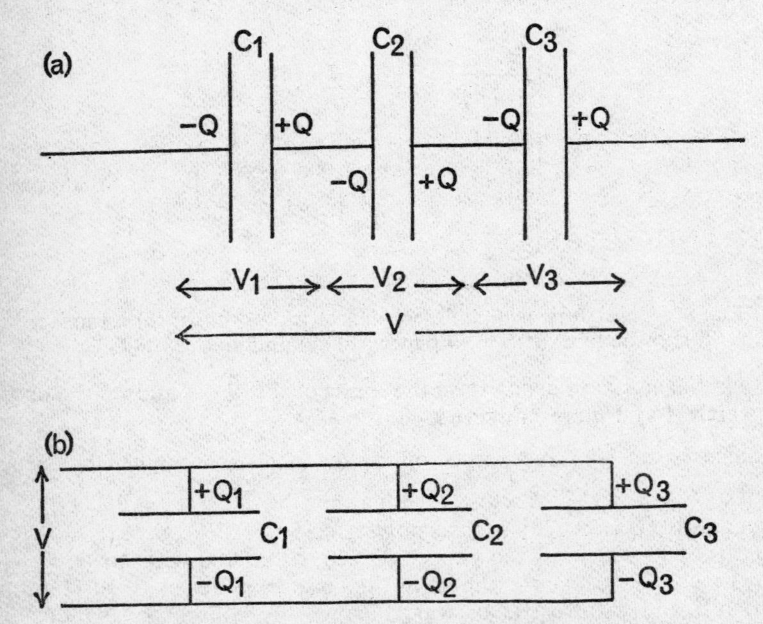

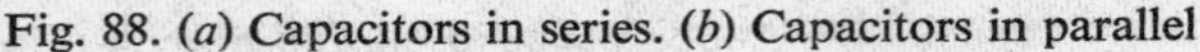
Fig. 88. (*a*) Capacitors in series. (*b*) Capacitors in parallel

equal number of electrons away from the second plate of C_1 (leaving it with a charge $+Q$) on to the first plate of C_2. Thus the first plate of C_2 acquires a charge $-Q$ which repels electrons away from the second plate of C_2 (leaving it with a charge $+Q$) on to the first plate of C_3. And so the chain reaction goes on: each capacitor ends up with a charge $-Q$ on its left-hand plate and $+Q$ on its right-hand plate.

Because the capacitors are connected in series, the p.d. across the whole system is the sum of the voltages across each capacitor. Hence

$$V = V_1 + V_2 + V_3$$

The total charge on the whole system is, as explained above, equal to the charge on each capacitor. Hence

$$Q = Q_1 = Q_2 = Q_3$$

Dividing the first equation by the second,

$$\frac{V}{Q} = \frac{V_1}{Q_1} + \frac{V_2}{Q_2} + \frac{V_3}{Q_3}$$

By definition, capacitance is the ratio of charge to p.d., or $C = \frac{Q}{V}$. We may therefore write $\frac{V}{Q} = \frac{1}{C}$, $\frac{V_1}{Q_1} = \frac{1}{C_1}$ etc., and the above equation becomes

$$\frac{1}{C} = \frac{1}{C_1} + \frac{1}{C_2} + \frac{1}{C_3} \qquad (1)$$

where C is the combined capacitance of the individual capacitors C_1, C_2, and C_3. The formula for capacitors in series (eqn. (1)) thus corresponds to the formula for resistors in parallel.

CAPACITORS IN PARALLEL

When capacitors are connected in parallel, as shown in Fig. 88 (*b*), the same p.d. exists across each of them. The total charge on the system is the sum of the charges on each capacitor, or

$$Q = Q_1 + Q_2 + Q_3$$

dividing by V (the p.d. across each)

$$\frac{Q}{V} = \frac{Q_1}{V} + \frac{Q_2}{V} + \frac{Q_3}{V}$$

Since, by definition, $C = \frac{Q}{V}$ we can write

$$C = C_1 + C_2 + C_3 \qquad (2)$$

where C is the combined capacitance of the individual capacitors

C_1, C_2, and C_3. The formula for capacitors in parallel (eqn. (2)) thus corresponds to the formula for resistors in series.

CAPACITORS IN A.C. CIRCUITS

Any capacitor offers an infinitely great resistance to direct current, since its dielectric is in theory a perfect insulator. But the situation *appears* to be different where alternating currents are concerned, because the current can be observed to be flowing on each side of the capacitor: but this is really the current that takes electrons to one plate and away from the other. The plates of a capacitor in an a.c. circuit become charged, discharged, then charged again with opposite polarity during each cycle of alternating e.m.f. Hence electrons surge to and fro in the

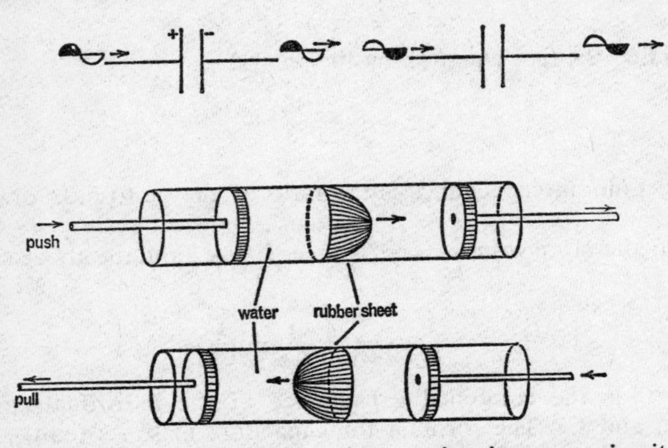

Fig. 89. Water-pipe analogy of a capacitor in an a.c. circuit

wires leading to the capacitor just as if they were passing through the dielectric.

This situation seems very curious at first sight, but it may be more easily understood in terms of the water-pipe analogy. A rubber sheet across a pipe (Fig. 89) prevents water flowing from one end to the other, just as the dielectric prevents electrons flowing through a capacitor. But since rubber is flexible, a to-and-fro motion applied to the water by a piston at one end of the pipe will cause a to-and-fro motion in a similar piston at the other end. The piston motion apparently passes through the rubber, even though the sheet is waterproof.

Similarly, a capacitor does not bar the to-and-fro motion of electrons, even though no electrons actually pass through the dielectric. Hence an alternating current can *in effect* pass through a capacitor, but direct current cannot.

The capacitor does, however, offer a considerable opposition to the passage of alternating current, and this opposition is called *capacitive reactance*. Being a special kind of resistance, capacitive reactance is measured in ohms. Its symbol is X_c.

X_c decreases as the frequency of the current is increased. A capacitor which offers considerable opposition to a 50-Hz current will probably provide an easy path for a radio-frequency current of 50 MHz. In the extreme case when the frequency is zero (i.e. a direct current) the capacitive reactance is infinitely great and no current can pass.

The fact that high-frequency currents pass easily through a capacitor is used in suppressing sources of radio interference. Sparking is always liable to occur between switch contacts and between commutators and brushes; this is not serious in itself, but it is troublesome because the high-frequency current oscillations set up can interfere with radio and television reception. A *small* capacitor connected across the switch contacts or from one contact to earth is virtually a non-conducting path for 50-Hz current, but provides a very easy path for any *sudden* voltage changes.

INDUCTORS

The characteristic property of a circuit that is responsible for the production of a back e.m.f. or induced voltage when there is a change of flux in the circuit is called inductance (see p. 113). An *inductor* is any component that is deliberately used because it possesses this property of inductance. In practice, an inductor, sometimes called a choke, is a coil; it may be wound on a hollow air-filled former or it may have a core of some magnetic material. Various types of inductor are shown in Fig. 90.

The inductance (strictly speaking, the coefficient of self-inductance) L of a coil is defined by the equation

$$E = -L\frac{\Delta I}{\Delta t}$$

where E is the voltage induced by the small change in current ΔI occurring in the small interval of time Δt (see p. 116).

INDUCTORS IN SERIES AND IN PARALLEL

The combined inductance of a number of inductors in series is simply the sum of the individual inductances, i.e.

$$L = L_1 + L_2 + L_3 +$$

(corresponding to the formula for resistors in series).

Similarly, the combined inductance of a number of inductors in parallel is given by

$$\frac{1}{L} = \frac{1}{L_1} + \frac{1}{L_2} + \frac{1}{L_3} +$$

(corresponding to the formula for resistors in parallel).

These formulae assume that there is no mutual inductance between the coils. If two series-connected inductors are so close together that their magnetic fields interlink their combined inductance is

either $L_1 + L_2 + 2M$

or $L_1 + L_2 - 2M$

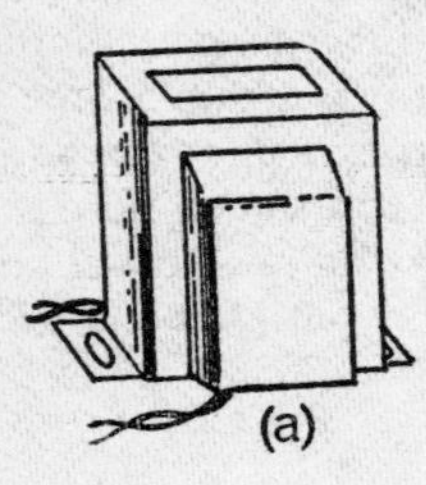

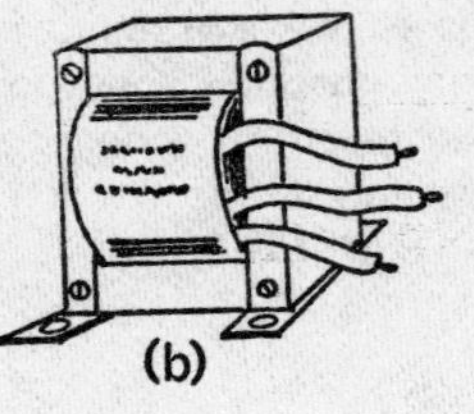

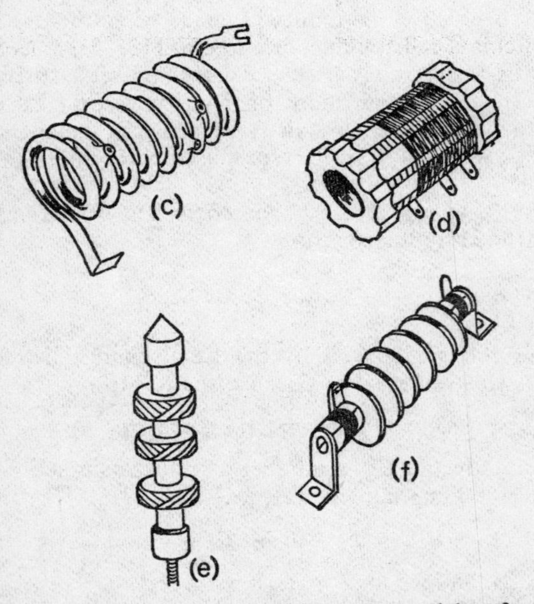

Fig. 90. Types of inductor: (*a*) and (*b*) iron-cored low-frequency chokes; (*c*) low-loss, air-cored, self-supporting coil; (*d*) single-layer, air-cored coil on ribbed frame; (*e*) and (*f*) multi-layer radio-frequency chokes wound in sections to reduce capacitive effects

where M is the coefficient of mutual inductance. The first alternative applies when current flows through both coils in the same direction and the fields assist each other (Fig. 91 (*a*)). The second alternative applies when the current flows in opposite directions in the two coils so that their magnetic fields are in opposition (Fig. 91 (*b*)).

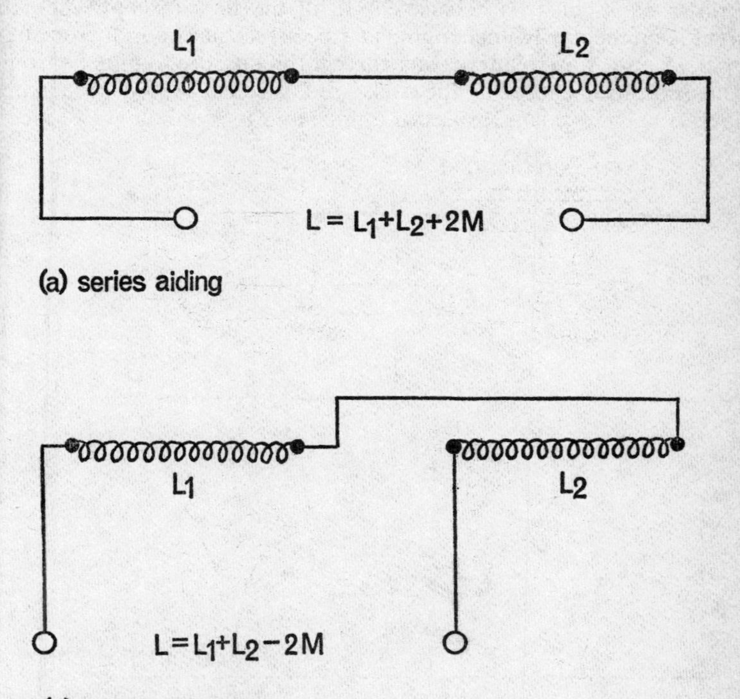

Fig. 91. Two methods of connecting inductors in series

EXAMPLE: A 10-henry coil is joined in series with a 20-henry coil. If the mutual inductance between them is 5 henrys, what are the two possible values of the combined inductance?

(*a*) Series aiding:

$$L = L_1 + L_2 + 2M$$
$$= 10 + 20 + 2 \times 5$$
$$= 40 \text{ henrys} \quad \textit{Answer}$$

(*b*) Series opposing:

$$L = L_1 + L_2 - 2M$$
$$= 10 + 20 - 2 \times 5$$
$$= 20 \text{ henrys} \quad \textit{Answer}$$

INDUCTIVE REACTANCE

When direct current begins to flow in an inductor a back e.m.f. is induced which opposes the current. But once the current has reached its steady value the back e.m.f. disappears. An alternating current, on the other hand, induces a back e.m.f. all the time (except when the current is momentarily unchanging at a peak). Hence an inductor offers a greater opposition to alternating current than it does to direct current.

The opposition caused by the back e.m.f. is called *inductive reactance*. Inductive reactance is measured in ohms, and its symbol is X_L.

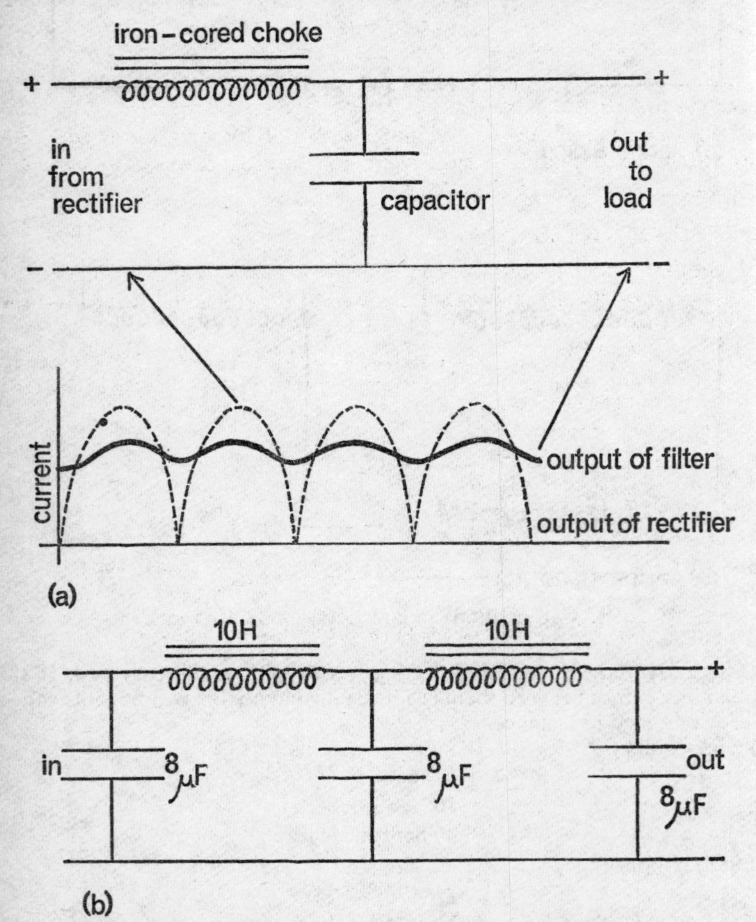

Fig. 92. Smoothing circuits: (*a*) simple choke-input filter; (*b*) practicable multi-section filter

X_L increases as the frequency of the current is increased. In the extreme case where the frequency is zero (i.e. a direct current) $X_L = 0$.

SMOOTHING CIRCUITS

The fact that low-frequency currents pass easily through an inductor is exploited in filter circuits used for smoothing the d.c. output of a rectifier (see p. 135). The rectified current is a pulsating direct current; it does not change direction, but its magnitude fluctuates with the frequency of the original alternating current (twice the frequency in the case of a full-wave rectifier).

An iron-cored inductor or choke connected in series with the output of the rectifier readily passes the direct current, but opposes the a.c. pulsations or 'ripple'. Any fluctuations in the current that remain after it has passed through the choke are largely short-circuited through a capacitor connected across the output of the filter (Fig. 92 (*a*)). The smoothing action can be improved by adding more inductors and capacitors, as shown in Fig. 92 (*b*).

CROSSOVER NETWORK

A similar capacitor–inductor combination forms a crossover network for separating the loudspeakers in a high-fidelity sound system. The

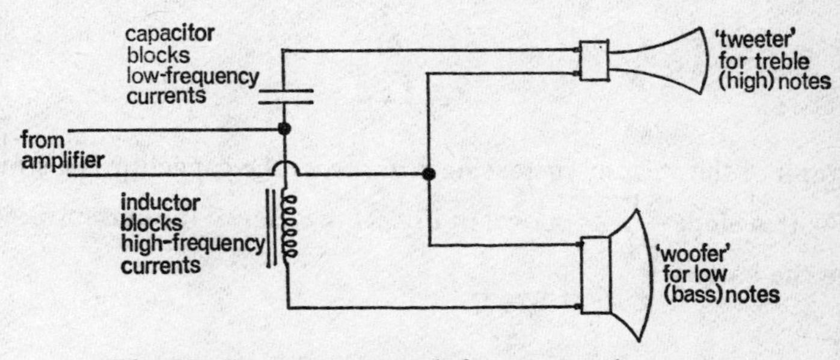

Fig. 93. Crossover network for two-speaker system

simplest 'hi-fi' installation employs two loudspeakers: a 'tweeter' for treble (high-frequency) notes and a 'woofer' for bass (low-frequency) notes. The output from the amplifier contains both high- and low-frequency currents, and the purpose of the crossover network is to split the output into a low-frequency component for the woofer and a high-frequency component for the tweeter.

A simple crossover network is shown in Fig. 93. It consists of a capacitor in series with the tweeter to block low-frequency currents, and a choke in series with the woofer to block high-frequency currents. The values of the capacitor and inductor are generally chosen to separate the frequencies at 3–5 kHz.

REACTANCE AND FREQUENCY

We have talked rather loosely so far of the connexion between frequency and the reactance of a capacitor or an inductor. Now we can investigate the exact relationship by applying mathematical reasoning to what we already know about capacitance and inductance.

X_c AND FREQUENCY

Capacitance is defined by the formula

$$C = \frac{Q}{V} \quad \text{(p. 143)}$$

or

$$Q = CV$$

When a capacitor is charged up steadily from a battery the charge increases by a small amount ΔQ (coulombs) in Δt sec; the potential difference between the plates increases by ΔV volts in the same time Δt, and

$$\Delta Q = C\Delta V$$

The current I amps drawn from the battery is $\frac{\Delta Q}{\Delta t}$ since current is the rate of flow of charge. Hence

$$I = \frac{\Delta Q}{\Delta t}$$

$$= \frac{C\Delta V}{\Delta t}$$

A graph of the steadily increasing p.d. across the capacitor is a straight line with a slope $\frac{\Delta V}{\Delta t}$, as shown in Fig. 94 (*a*). Hence the current flowing from the battery is

$$I = \frac{C\Delta V}{\Delta t}$$

$$= C \times \text{slope of } V/t \text{ graph}$$

If the battery is replaced by an a.c. supply the potential difference across the capacitor varies with time in the same way as the supply voltage, which for simplicity we will assume to be

$$E = E_{\max} \sin \theta$$

So the p.d. across a capacitor connected to an alternating supply varies with time according to the formula

$$V = V_{\max} \sin \theta$$

and a graph of this p.d. is not a straight line, as it was when the capacitor was charged from a battery, but a sine wave (Fig. 94 (*b*)).

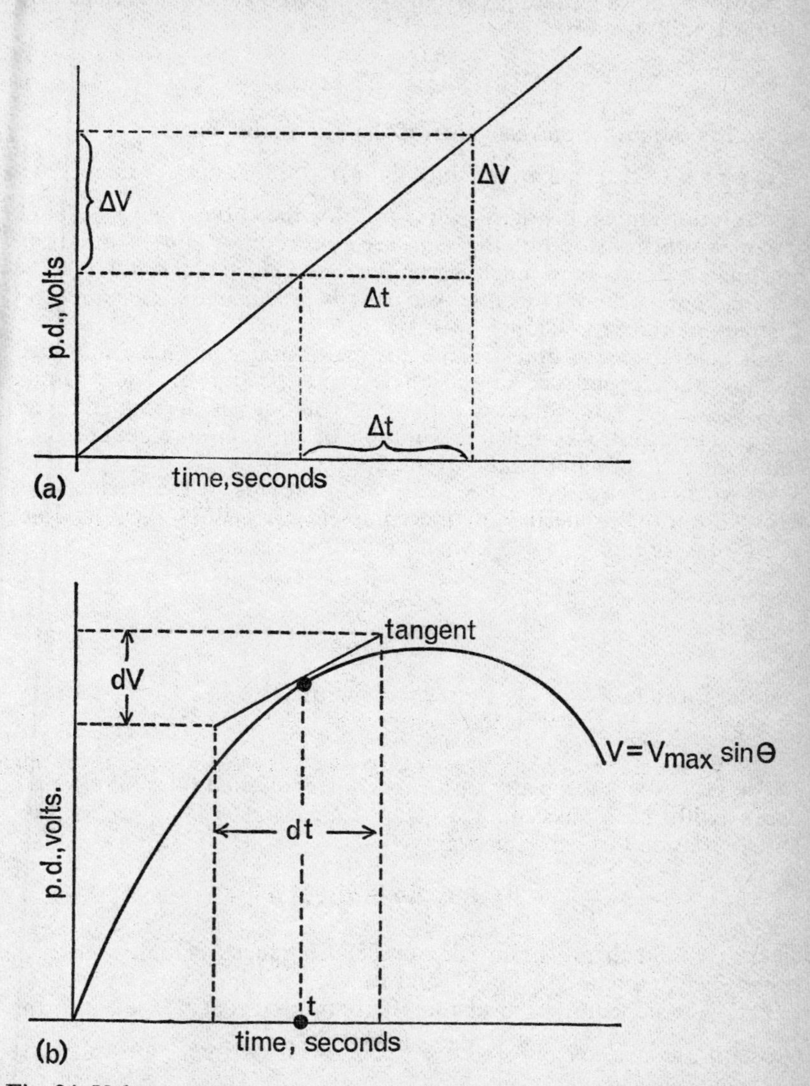

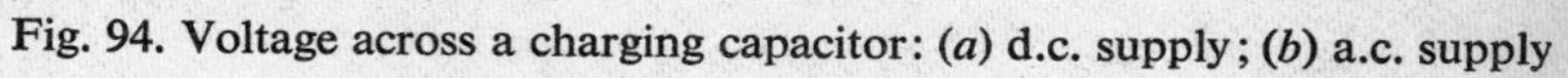
Fig. 94. Voltage across a charging capacitor: (*a*) d.c. supply; (*b*) a.c. supply

However, the alternating current flowing into the capacitor at any instant is still given by

$$i = \frac{\Delta Q}{\Delta t} = C\frac{\Delta V}{\Delta t}$$

where i is the *instantaneous* value of the alternating current. The difficulty now is to determine $\frac{\Delta V}{\Delta t}$, since the V/t graph is not a straight line. One solution shown in Fig. 94 (*b*) is to draw the curve $V = V_{max} \sin \theta$ and measure the slope of the *tangent* to it at the point in question. Fortunately there is a much better method which does not involve drawing curves and tangents: we can use elementary calculus and *differentiate* the expression $V = V_{max} \sin \theta$.

But before we can differentiate this expression it must be modified to show the dependence of the angle θ on the time t. The formula $E = E_{max} \sin \theta$ was obtained for a coil rotating in a magnetic field (see p. 110), and a complete revolution of the coil corresponds to a complete cycle of e.m.f. and current.

For one complete cycle, $\theta = 360°$ or 2π radians. If the frequency is f, the duration or period of one complete cycle is $1/f$ sec. In other words, θ is 2π radians (360°) when t is $1/f$ sec. Hence

$$\frac{\theta}{2\pi} = \frac{t}{1/f}$$

or

$$\theta = 2\pi ft$$

and the formula $E = E_{max} \sin \theta$ can be written as

$$E = E_{max} \sin 2\pi ft$$

The p.d. across the plates of a capacitor connected to an a.c. supply varies with time according to the expression $V = V_{max} \sin 2\pi ft$. Differentiating this expression gives

$$\frac{dV}{dt} = V_{max} \times 2\pi f \times \cos 2\pi ft$$

where dV and dt are infinitesimally small changes of p.d. and time, respectively, compared with ΔV and Δt.

The instantaneous value of the alternating current flowing into the capacitor, shown above to be $i = C\frac{\Delta V}{\Delta t}$, can be written more correctly as

$$i = C\frac{dV}{dt}$$
$$= C \times V_{max} \times 2\pi f \times \cos 2\pi ft$$
$$= 2\pi fCV_{max} \cos 2\pi ft \qquad (1)$$

The maximum value of eqn. (1) occurs when $\cos 2\pi ft = 1$ (since cosines, like sines, vary between 0 and 1), hence the maximum current (I_{max}) is found by putting $\cos 2\pi ft = 1$ in eqn. (1)

$$I_{max} = 2\pi fCV_{max}$$

Furthermore, since the effective or r.m.s. current $I = I_{max}/\sqrt{2}$ (p. 134) and the effective or r.m.s. voltage $V = V_{max}/\sqrt{2}$, we can write

$$I = 2\pi fCV \qquad (2)$$

where I and V are r.m.s. values of current and voltage.

Just as the ratio of voltage to current in a d.c. circuit defines the resistance or opposition to current, so in a purely capacitive circuit (one having zero resistance and zero inductance) the ratio V/I defines the *capacitive reactance* X_c.

Rearranging eqn. (2),

$$\frac{V}{I} = \frac{1}{2\pi fC}$$

Hence the relationship between capacitive reactance and frequency is

$$X_c = \frac{1}{2\pi fC} \qquad (3)$$

EXAMPLE: A 100-pF capacitor is connected across an oscillator producing 20 V at a frequency of 5 megahertz (5×10^6 c/s). What is the capacitive reactance and the current drawn from the oscillator?

$$V = 20 \text{ V}$$
$$f = 5 \times 10^6$$
$$C = 100 \times 10^{-12} \text{ F}$$
$$X_c = \frac{1}{2\pi fC}$$
$$= \frac{1}{2 \times 3{\cdot}142 \times 5 \times 10^6 \times 100 \times 10^{-12}}$$
$$= 318 \text{ ohms} \quad \textit{Answer}$$
$$X_c = V/I$$

hence

$$I = V/X_c$$
$$= 20/318$$
$$= 0{\cdot}063 \text{ amps or } 63 \text{ mA} \quad \textit{Answer}$$

X_L AND FREQUENCY

In a purely inductive circuit the ratio of voltage to current defines the *inductive reactance* X_L.

When an alternating voltage V is applied to an inductor a back e.m.f.

is induced whose magnitude is proportional to the rate of change of current, i.e.

$$E = -L\frac{\Delta I}{\Delta t}$$

(see p. 117). Assuming that the circuit has zero resistance and zero capacitance, the back e.m.f. must be exactly equal and opposite to the applied voltage. Hence we may write

$$V = L\frac{\Delta I}{\Delta t}$$

or more precisely, using instantaneous values,

$$v = L\frac{\mathrm{d}i}{\mathrm{d}t}$$

where $\mathrm{d}i$ and $\mathrm{d}t$ are infinitesimally small compared with ΔI and Δt.

If the alternating current is of the type $i = I_{max} \sin \theta$ we know that the relationship between current and time at any instant is

$$i = I_{max} \sin 2\pi ft$$

from which the rate of change of current can be found by the process of differentiation. Differentiating $i = I_{max} \sin 2\pi ft$ gives

$$\frac{\mathrm{d}i}{\mathrm{d}t} = I_{max} \times 2\pi f \times \cos 2\pi ft$$

and the instantaneous value of the applied voltage can now be written as

$$v = L\frac{\mathrm{d}i}{\mathrm{d}t} = 2\pi fLI_{max} \cos 2\pi ft \qquad (4)$$

The maximum voltage occurs when $\cos 2\pi ft = 1$, hence

$$V_{max} = 2\pi fLI_{max}$$

Furthermore, since $V_{max} = \sqrt{2}V$ and $I_{max} = \sqrt{2}I$, where V and I are the effective or r.m.s. values of voltage and current, respectively,

$$\sqrt{2}V = 2\pi fL\sqrt{2}I$$

or

$$V = 2\pi fLI \qquad (5)$$

rearranging eqn. (5) gives the ratio of voltage to current

$$\frac{V}{I} = 2\pi fL$$

and this ratio is by definition the *inductive reactance* X_L. Hence the relationship between inductive reactance and frequency is

$$X_L = 2\pi fL \qquad (6)$$

EXAMPLE: A 2-H inductance coil is connected across a 100-V 50-Hz supply. What is the inductive reactance? Neglecting any resistance, what current flows in the coil?

$$V = 100 \text{ V}$$
$$f = 50$$
$$L = 2 \text{ H}$$
$$X_L = 2\pi fL$$
$$= 2 \times 3{\cdot}142 \times 50 \times 2$$
$$= 628 \text{ ohms} \quad \textit{Answer}$$

$$X_L = V/I$$
$$I = V/X_L$$
$$= 100/628$$
$$= 0{\cdot}159 \text{ amp} \quad \textit{Answer}$$

Summarizing the results:
when an alternating voltage

$$v = V_{max} \sin 2\pi ft$$

is applied to a capacitor the instantaneous value of the current is (eqn. (1))

$$i = 2\pi fCV_{max} \cos 2\pi ft$$
$$= \left(\frac{V_{max}}{Xc}\right) \cos 2\pi ft$$
$$= I_{max} \cos 2\pi ft$$

The current and voltage waveforms are shown in Fig. 95.

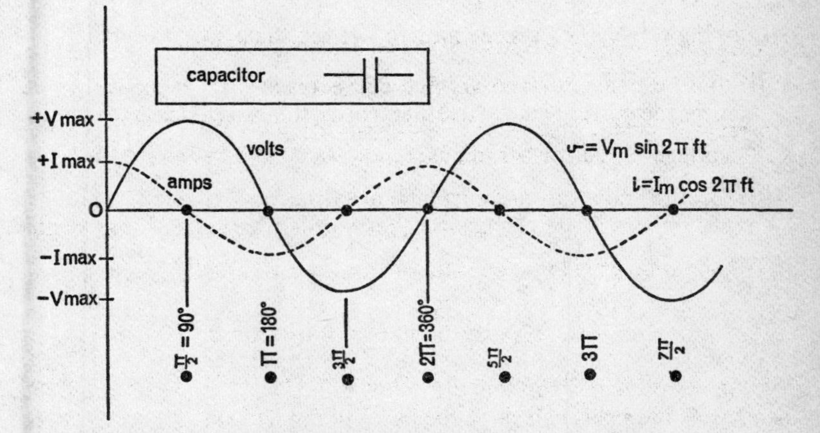

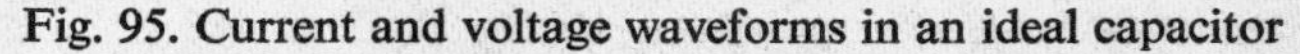
Fig. 95. Current and voltage waveforms in an ideal capacitor

When an alternating current

$$i = I_{\max} \sin 2\pi ft$$

flows in an inductor the instantaneous value of the p.d. across its ends is (eqn. (4))

$$\begin{aligned} v &= 2\pi f L I_{\max} \cos 2\pi ft \\ &= (X_L I_{\max}) \cos 2\pi ft \\ &= V_{\max} \cos 2\pi ft \end{aligned}$$

The current and voltage waveforms are shown in Fig. 96.

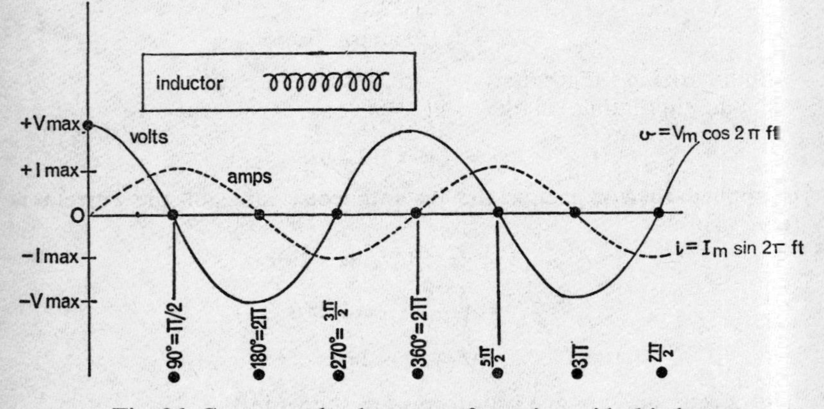

Fig. 96. Current and voltage waveforms in an ideal inductor

Looking closely at Figs. 95 and 96, we see that:

(i) for the capacitor the voltage curve reaches its peak values 90° after the current curve; in other words, the current *leads* the applied voltage by 90° or $\frac{\pi}{2}$ radians;

(ii) for the inductor the current curve reaches its peak values 90° after the voltage curve; in other words, the current *lags* behind the voltage by 90° or $\frac{\pi}{2}$ radians.

For both the capacitor and the inductor we can say that the current and voltage are out of step or, to use the proper term, *out of phase* by 90°. For a pure resistor, on the other hand, current and voltage are 'in phase'; the peak values of the current and voltage occurring at the same instant.

A surprising result is obtained when we calculate the *power* in a pure inductor or a pure capacitor. Power is the product of current and voltage ($P = IV$), and the power at any instant is found by multiplying the instantaneous current by the instantaneous voltage. If the voltage and current are in phase, as they are in a pure resistor, the power is never negative (see Fig. 76), since the product $(+i) \times (+v)$ is positive and the product $(-i) \times (-v)$ is also positive.

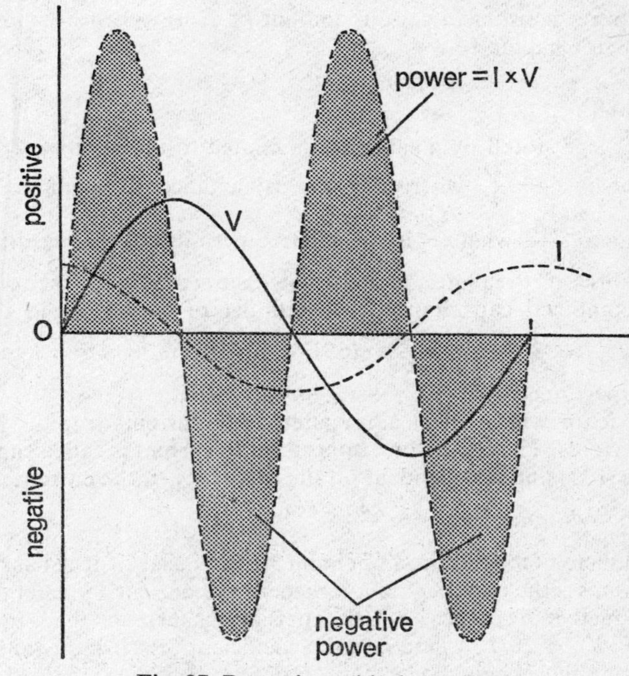

Fig. 97. Power in an ideal capacitor

But in the case of a capacitor or an inductor, current and voltage are out of phase by 90°. There are times when the current is negative and the voltage is positive, and the product $(-i) \times (+v)$ is negative, i.e. the power is negative, as shown in Fig. 97. The power is also negative at times when current is positive and voltage is negative, since the product $(+i) \times (-v)$ is negative.

Fig. 97 is the power curve for an ideal capacitor—a very similar curve is obtained for the power in an ideal inductor—and it has two positive and two negative lobes for each complete voltage cycle. Since positive power means energy consumed or absorbed by the circuit, while negative power means energy returned from the circuit, it is clear that the net consumption of energy is nil. In other words, no real energy is consumed in an ideal capacitor—or in an ideal inductor.

During one quarter-cycle energy is supplied to the capacitor (or inductor), and in the next quarter-cycle the same amount of energy is returned to the generator. Physically, this means that all the energy stored in the electric field of a capacitor (or in the magnetic field of an inductor) during charging is released during discharge.

But all actual capacitors and inductors have some resistance, and hence always consume a certain amount of energy. Just how much we shall see in Chapter Ten.

SUMMARY

The charge stored by a capacitor is related to the voltage across the capacitor by $C = \frac{Q}{V}$, where C is the capacitance. Capacitance is proportional to $\frac{\varepsilon_r A}{d}$, where ε_r is the relative permittivity of the dielectric, A is the area of the plates, and d is the distance between them.

The combined capacitance of a number of capacitors in series is given by $\frac{1}{C} = \frac{1}{C_1} + \frac{1}{C_2} + \frac{1}{C_3}$, etc. For capacitors in parallel the combined capacitance is $C = C_1 + C_2 + C_3$.

A capacitor will in effect allow alternating currents to pass, but not direct currents. The capacitor's opposition to current is called capacitive reactance (X_c) and is dependent on the frequency of the current according to the relationship $X_c = \frac{1}{2\pi fC}$.

An inductor (coil) in an a.c. circuit has induced in it a back e.m.f. which opposes the current. Inductors therefore present a greater opposition (inductive reactance X_L) to alternating currents than to direct currents: $X_L = 2\pi fL$, where L is the coefficient of self-inductance.

The combined inductance of a number of inductors in series is $L = L_1 + L_2 + L_3$. The combined inductance of a number of inductors in parallel is given by $\frac{1}{L} = \frac{1}{L_1} + \frac{1}{L_2} + \frac{1}{L_3}$. If two inductors in series are linked by mutual inductance M the combined inductance is

$$L = L_1 + L_2 \pm 2M$$

When an alternating voltage $v = V_{max} \sin 2\pi ft$ is applied to a capacitor the instantaneous value of the current is

$$i = I_{max} \cos 2\pi ft$$

The current leads the voltage by 90°.

When an alternating current $i = I_{max} \sin 2\pi ft$ flows in an inductor the instantaneous value of the voltage is

$$v = V_{max} \cos 2\pi ft$$

The current lags behind the voltage by 90°.

The power in an inductor or capacitor is positive for half the time and negative for the remainder, hence there is no net consumption of energy in a purely capacitive circuit or in a purely inductive circuit.

CHAPTER TEN

ALTERNATING-CURRENT CIRCUITS

Because alternating currents and voltages go through the same cyclic pattern over and over again, we do not bother to follow their instantaneous changes—we prefer to deal instead with their r.m.s. values. But in doing so we have lost touch with their phase angles, and we therefore cannot just add r.m.s. values together. However, we can put

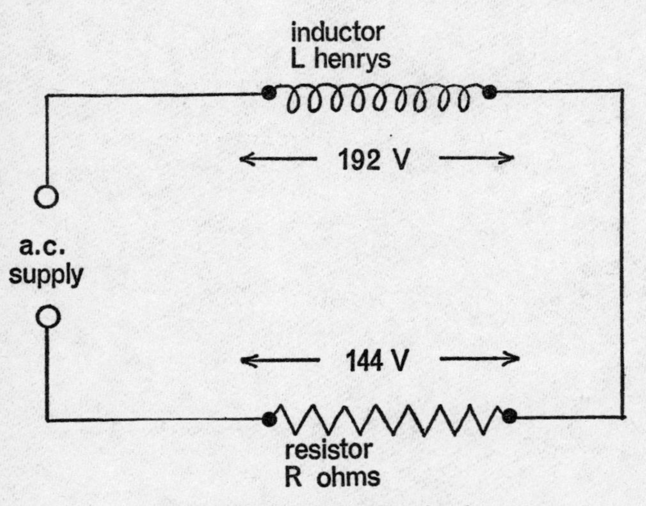

Fig. 98. Simple inductive circuit

the angles back again by a simple and elegant graphical method. The procedure is to treat them as *vectors*, i.e. to represent the magnitude of each electrical quantity by the *length* of a line whose *direction* corresponds to the phase relationship of the quantity in question.

As an example, let us analyse the circuit shown in Fig. 98, consisting of an inductor and a resistor connected to an a.c. supply. A voltmeter reads 192 V (r.m.s.) across the inductor and 144 V across the resistor.

We know that the current and voltage in the resistor are 'in phase', i.e. the phase angle between them is 0°. Since angles are normally measured from the horizontal, we can draw a horizontal line to represent the resistor voltage; its length is to be proportional to 144 V.

We also know (see p. 160) that the current in the inductor lags 90° behind the voltage, i.e. the voltage is 90° *ahead* of the current. We can therefore draw a line at 90° to the horizontal to represent the inductor voltage, as shown in Fig. 99. Its length is to be proportional to 192 V.

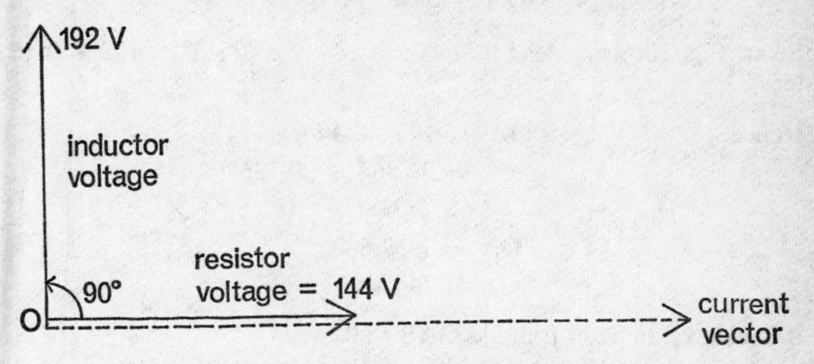

Fig. 99. Vector diagram of voltages in resistor and inductor

The requirement now is to find the sum of these two voltage vectors. Finding the sum of two or more vectors is a fairly common geometrical problem and is solved by placing the vectors toe to tip and drawing the *resultant* from the starting-point of the first vector to the end point

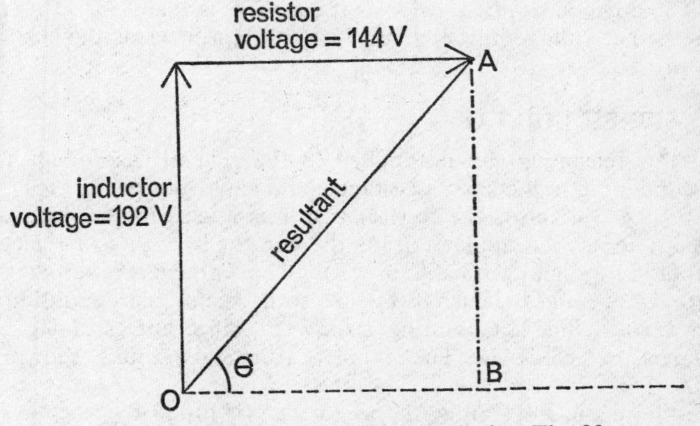

Fig. 100. Vector addition applied to Fig. 99

of the last vector. Fig. 100 shows this method of vector addition applied to the voltage diagram of Fig. 99.

The resultant is OA, and its direction (the angle it makes with the horizontal) is $A\hat{O}B = \theta$. Of course the length of OA and the size of the angle θ could both be found from a scale drawing, but it is more accurate to calculate them as follows.

OAB is a right-angled triangle, hence by Pythagoras' theorem

$$OA^2 = AB^2 + OB^2$$

From Fig. 100 it is clear that OB = 144 V and AB = 192 V (r.m.s. values).

Hence
$$\begin{aligned} OA^2 &= 192^2 + 144^2 \\ &= 36{,}864 + 20{,}736 \\ &= 57{,}600 \\ OA &= \sqrt{57{,}600} \\ &= 240 \text{ V} \end{aligned}$$

It is also clear from triangle OAB that

$$\begin{aligned} \tan \theta &= \frac{AB}{OB} \\ &= \frac{192}{144} \\ &= 1{\cdot}333 \end{aligned}$$

From tables of tangents we find that the angle whose tangent is 1·333 is 53°. The angle θ, or *phase angle* as it is called, is therefore 53°.

Thus the net voltage in the circuit is 240 V, and it leads the current by 53°.

R–L–C SERIES CIRCUIT

The same reasoning can be applied to the general case of an a.c. circuit containing inductance, resistance, and capacitance (Fig. 101).

The voltage V_R across the resistor is in phase with the current; the voltage V_L across the inductor leads the current by 90°, so its vector representation points vertically upward. The voltage V_c across the capacitor *lags behind* the current by 90°, so its vector representation is another vertical line but pointing downward, since angles of lag are usually measured clockwise. The complete voltage diagram is shown in Fig. 102.

As in the previous example, the vector sum is found by placing the three voltage vectors toe to tip and drawing the resultant from the start of the first to the end of the last (Fig. 103).

Because OAB is a right-angled triangle

$$\begin{aligned} OA^2 &= OB^2 + AB^2 \\ OA &= \sqrt{OB^2 + AB^2} \end{aligned}$$

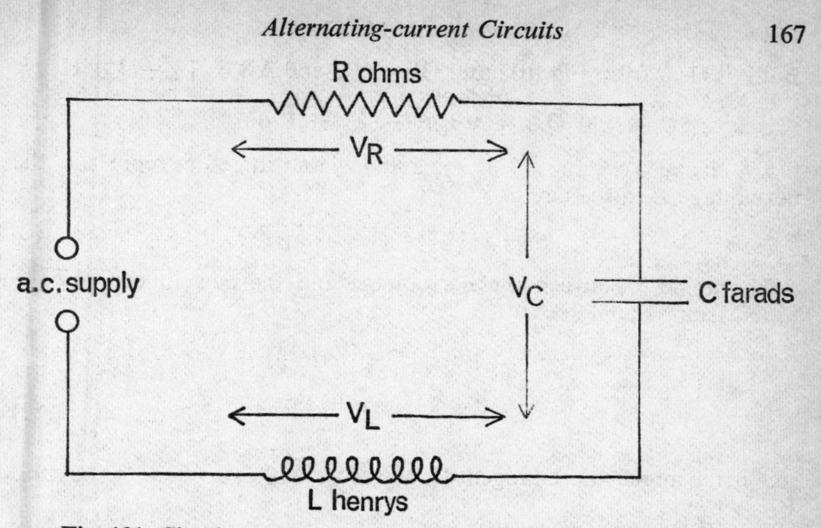

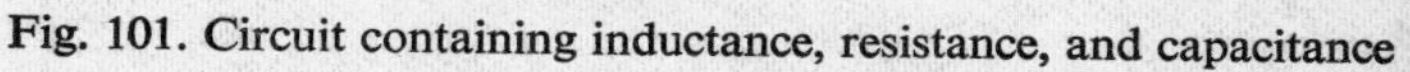
Fig. 101. Circuit containing inductance, resistance, and capacitance

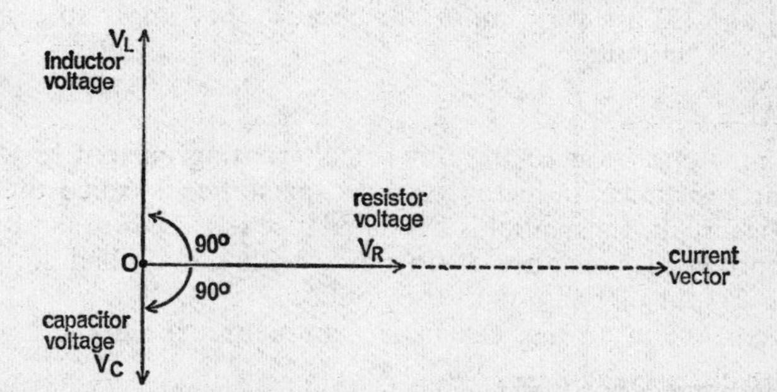

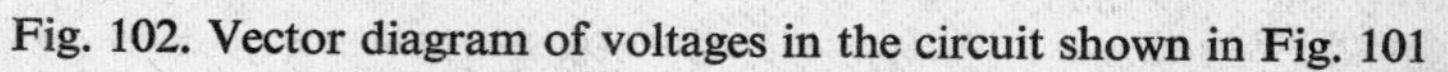
Fig. 102. Vector diagram of voltages in the circuit shown in Fig. 101

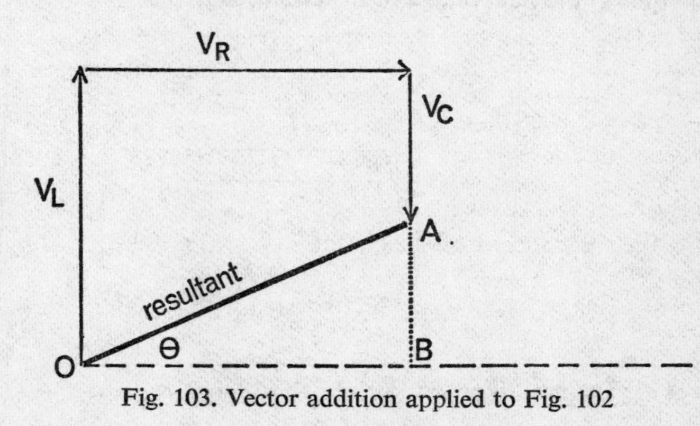

Fig. 103. Vector addition applied to Fig. 102

But it is clear from Fig. 103 that OB $= V_R$ and AB $= V_L - V_C$. Hence

$$OA = \sqrt{V_R^2 + (V_L - V_C)^2}$$

OA is the vector sum of V_R, V_L, and V_C, and therefore represents the net voltage E in the circuit

$$E = \sqrt{V_R^2 + (V_L - V_C)^2} \quad (1)$$

The angle θ between the net voltage and the current can also be found from triangle OAB:

$$\tan \theta = \frac{AB}{OB}$$

$$= \frac{V_L - V_C}{V_R} \quad (2)$$

so θ is the angle whose tangent (found from tables) is equal to the ratio $\frac{V_L - V_C}{V_R}$.

Eqns. (1) and (2) are very important because they apply to *any* series-connected a.c. circuit.

IMPEDANCE

The *total* opposition to the flow of alternating current in a circuit containing resistance, inductance, and capacitance is called the *impedance* of the circuit. Its symbol is Z. Impedance is a vector quantity, since it is composed of resistance R (in phase with the applied e.m.f.) and reactance X (out of phase with the applied e.m.f.).

If the current in an a.c. circuit is I amps and the applied e.m.f. is E volts the impedance is $Z = \frac{E}{I}$.

From eqn. (1) the net voltage in the circuit is

$$E = \sqrt{V_R^2 + (V_L - V_C)^2}$$

hence

$$Z = \frac{E}{I} = \frac{\sqrt{V_R^2 + (V_L - V_C)^2}}{I}$$

$$= \sqrt{\frac{V_R^2}{I^2} + \left(\frac{V_L}{I} - \frac{V_C}{I}\right)^2}$$

but $\frac{V_R}{I}$ is the *resistance* R of the circuit

and $\frac{V_L}{I}$ is the *inductive reactance* X_L

and $\frac{V_C}{I}$ is the *capacitive reactance* X_C

Therefore $$Z = \sqrt{R^2 + (X_L - X_C)^2} \tag{3}$$

and this can be represented as a right-angled triangle, as shown in Fig. 104.

The phase angle θ between the resistance and the impedance is, of course, the same as the angle θ between the current and net voltage. From Fig. 104 it will be seen that

$$\sin\theta = \frac{X_L - X_C}{Z}$$

$$\cos\theta = \frac{R}{Z}$$

$$\tan\theta = \frac{X_L - X_C}{R}$$

and any of these relationships can be used in conjunction with trigonometrical tables to find θ.

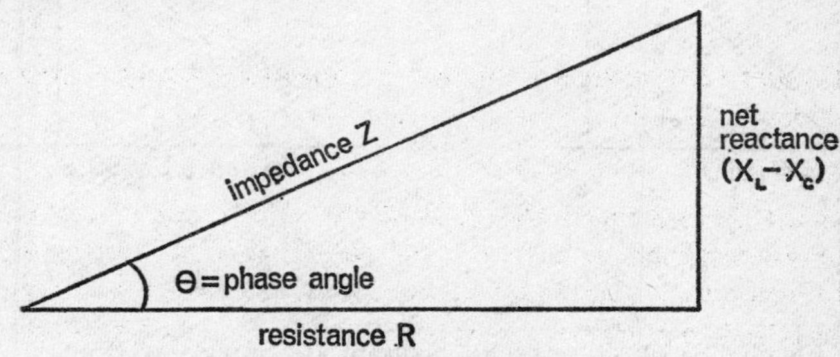

Fig. 104. Impedance triangle

EXAMPLE: A series a.c. circuit has a resistance of 90 Ω, an inductive reactance of 200 Ω, and a capacitive reactance of 80 Ω. Voltmeters connected across the components read 180 V across the resistor, 400 V across the inductor, and 160 V across the capacitor. What is the applied e.m.f., the impedance of the circuit, the phase angle, and the current?

First, let us construct the voltage vector diagram, as shown in Fig. 105 (*a*), from $V_R = 180$ V, $V_L = 400$ V, and $V_C = 160$ V.

The applied e.m.f. is given by eqn. (1):

$$\begin{aligned} E &= \sqrt{V_R^2 + (V_L - V_C)^2} \\ &= \sqrt{180^2 + (400 - 160)^2} \\ &= \sqrt{90{,}000} \\ &= 300 \text{ V} \quad \textit{Answer} \end{aligned}$$

The impedance triangle (Fig. 105 (*b*)) can be constructed from $R = 90$ Ω, $X_L = 200$ Ω, and $X_C = 80$ Ω.

The impedance is given by eqn. (3):

$$
\begin{aligned}
Z &= \sqrt{R^2 + (X_L - X_C)^2} \\
&= \sqrt{90^2 + (200 - 80)^2} \\
&= \sqrt{22{,}500} \\
&= 150\ \Omega \quad \textit{Answer}
\end{aligned}
$$

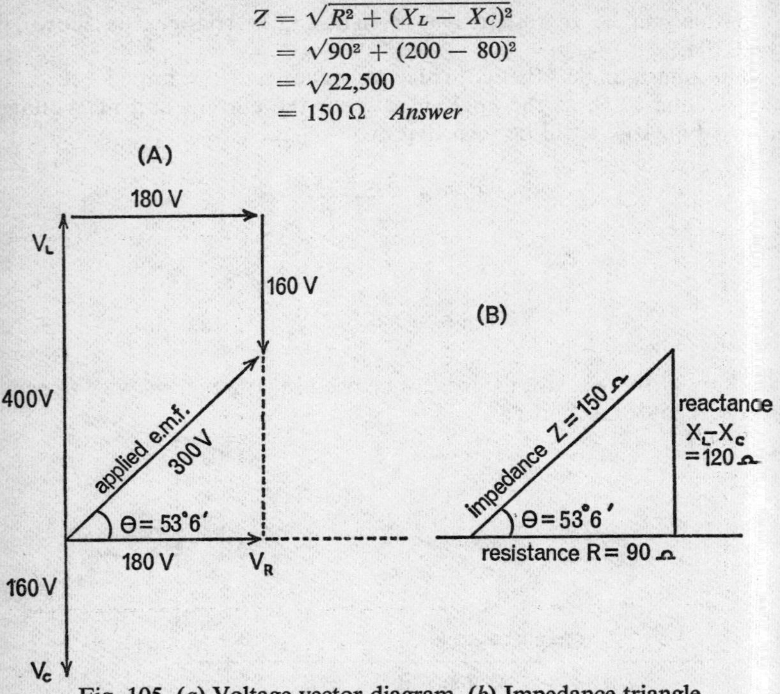

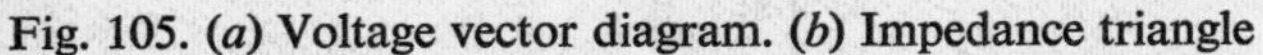

Fig. 105. (*a*) Voltage vector diagram. (*b*) Impedance triangle

The phase angle θ can be obtained either from the voltage vector diagram or from the impedance triangle. Using the latter,

$$
\begin{aligned}
\tan\theta &= \frac{X_L - X_C}{R} \\
&= \frac{200 - 80}{90} \\
&= 1{\cdot}333
\end{aligned}
$$

From tables, the angle whose tangent equals 1·333 is found to be 53°. *Answer*

Finally, the current is found from the formula

$$Z = \frac{E}{I}$$

or

$$
\begin{aligned}
I &= \frac{E}{Z} \\
&= \tfrac{300}{150} \\
&= 2 \text{ amps} \quad \textit{Answer}
\end{aligned}
$$

PARALLEL CIRCUITS

When components are connected in parallel (Fig. 106) the potential difference across each branch of the circuit is the same and is equal to the applied voltage. The sum of the branch currents is equal to the total current drawn from the source, but because of the different phase angles of the alternating currents, it is necessary to compute the *vector* sum, not the arithmetic sum.

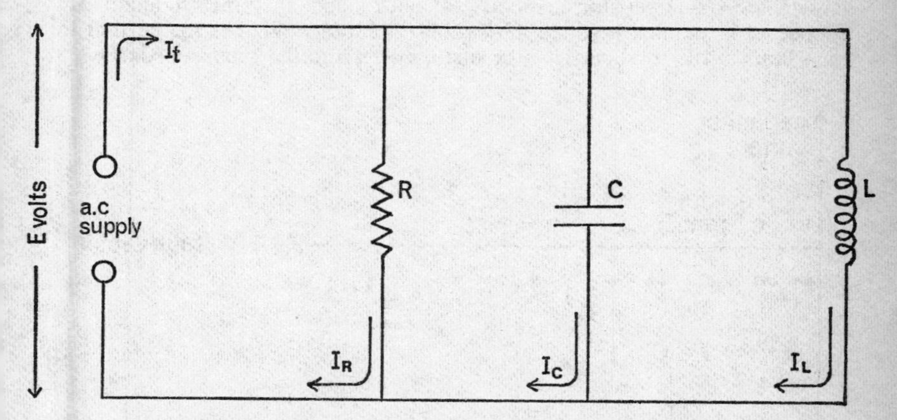

Fig. 106. Alternating current in a parallel-connected circuit

The current flowing through the resistor in Fig. 106 is $I_R = \frac{E}{R}$ amps. The reactance of the capacitor is X_C and the current flowing through it is $I_C = \frac{E}{X_C}$ amps. The reactance of the inductor is X_L, and the current flowing through it is $I_L = \frac{E}{X_L}$ amps. The total current drawn from the source is I_t amps.

In constructing the vector diagram of the various currents, we choose as our horizontal reference the e.m.f. of the source, since this voltage is common to all branches of the circuit. The resistor current I_R is in phase with the voltage, and is therefore represented by a horizontal vector. The capacitor current I_C *leads* the voltage by 90°, and is therefore represented by a vector pointing vertically upwards (Fig. 107). The inductor current I_L lags *behind* the voltage by 90°, and is therefore represented by a vector pointing vertically downward.

The vector sum shown in Fig. 108 represents the total current I_t. From the geometry of Fig. 108 we see that

$$I_t = \sqrt{I_R^2 + (I_L - I_C)^2}$$

and $$\tan\theta = \frac{I_L - I_C}{I_R}$$

where θ is the phase angle between current and voltage. (In this particular circuit it is evident that the current lags behind the voltage.) The total impedance of the circuit is $Z = \frac{E}{I_t}$.

EXAMPLE: A 6-Ω resistor, a 66-μF capacitor, and a 19-mH inductor are connected in parallel across a 48-V, 100-Hz supply. What is the current in each branch, the total current, the phase angle, and the total impedance?

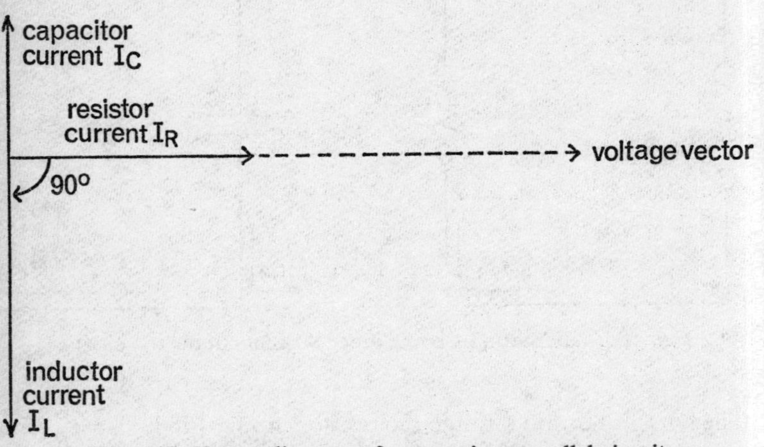

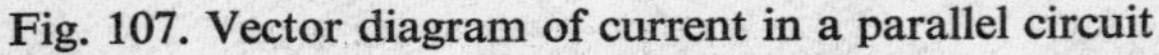

Fig. 107. Vector diagram of current in a parallel circuit

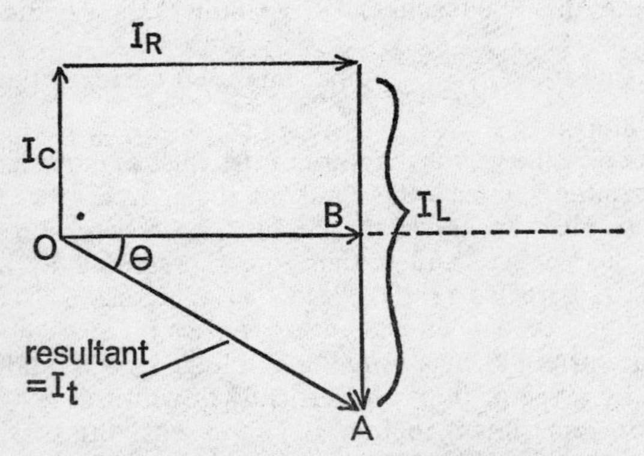

Fig. 108. Vector addition applied to Fig. 107

The current in the resistor is

$$I_r = \frac{E}{R}$$
$$= \tfrac{48}{6} = 8 \text{ amps} \quad \textit{Answer}$$

The reactance of the capacitor is

$$X_C = \frac{1}{2\pi fC}$$
$$= \frac{1}{2 \times 3{\cdot}1416 \times 100 \times 66 \times 10^{-6}}$$
$$= 24\ \Omega$$

and the current in the capacitor is

$$I_C = \frac{E}{X_C}$$
$$= \tfrac{48}{24} = 2 \text{ amps} \quad \textit{Answer}$$

The reactance of the inductor is

$$X_L = 2\pi fL$$
$$= 2 \times 3{\cdot}1416 \times 100 \times 19 \times 10^{-3}$$
$$= 12\ \Omega$$

and the current in the inductor is

$$I_L = \frac{E}{X_L}$$
$$= \tfrac{48}{12} = 4 \text{ amps} \quad \textit{Answer}$$

The total current is

$$I_t = \sqrt{I_R^2 + (I_L - I_C)^2}$$
$$= \sqrt{8^2 + (4 - 2)^2}$$
$$= \sqrt{68} = 8{\cdot}25 \text{ amps} \quad \textit{Answer}$$

The phase angle is given by

$$\tan\theta = \frac{I_L - I_C}{R}$$
$$= \frac{4 - 2}{8} = 0{\cdot}25$$

From tables it is found that $\tan 14° = 0{\cdot}25$. Hence

$$\theta = 14° \quad \textit{Answer}$$

Finally, the total impedance of the circuit is

$$Z = \frac{E}{I_t}$$

$$= \frac{48}{8{\cdot}25} = 5{\cdot}82\ \Omega \quad \textit{Answer}$$

POWER IN A.C. CIRCUITS

We saw in Chapter Nine that the power in an ideal inductor or capacitor consists of equal positive and negative lobes, so that the net power consumed in these components is nil. But all real inductors and

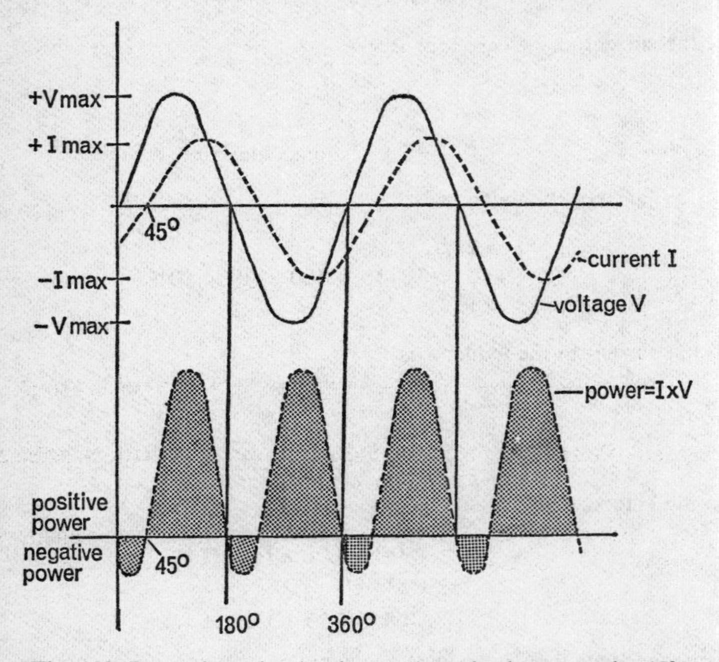

Fig. 109. Power in a circuit where voltage leads current by 45°

capacitors have some resistance. As a result, the current is out of phase with the voltage by an angle of less than 90°; although the power (current × voltage) still consists of positive and negative lobes, the positive lobes are larger than the negative. (This is illustrated in Fig. 109 for a phase angle of 45°.) The power consumed by the circuit is the difference between the positive and negative lobes.

Plotting the instantaneous current and voltage curves is a very tedious method of calculating the power consumed, so it is perhaps as well that the same result can be obtained from a voltage vector diagram.

We know that power is consumed only in the resistive component of the circuit. Hence the product of the current I_R and the voltage V_R in the *resistive* component gives the real power consumption of the *whole* circuit. Using symbols

$$P = I_R V_R \text{ watts}$$

where I_R is the effective (r.m.s.) value of the current in the resistive component, V_R is the effective (r.m.s.) value of the voltage across the resistive component, and P is the real power consumed by the circuit.

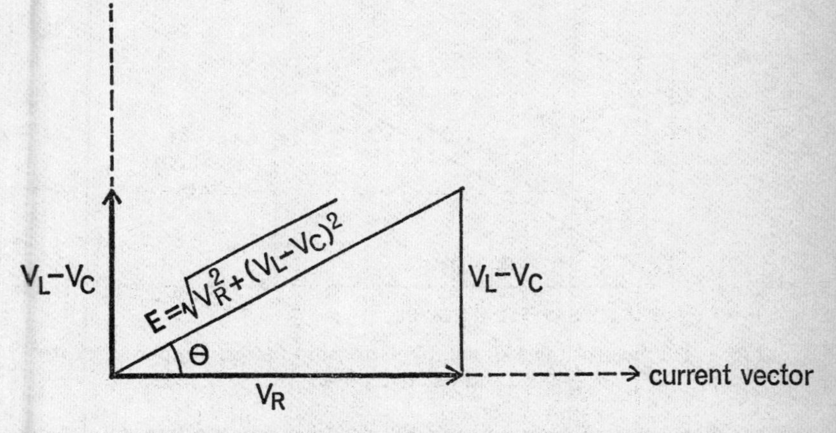

Fig. 110. Voltage vector diagram of an a.c. circuit containing resistance, inductance, and capacitance in series

For a series-connected circuit we also know from the voltage vector diagram (Fig. 110) that

$$\frac{V_R}{E} = \cos\theta$$

or

$$V_R = E\cos\theta$$

where E is the applied e.m.f. and θ is the phase angle.

Hence $$P = EI\cos\theta \quad \text{watts}$$

This is the power actually consumed in an a.c. circuit, and is known as the *real power*. The *apparent power*, i.e. the power which the circuit seems to be drawing from the a.c. supply, is the product of the current and the applied e.m.f. Thus if a voltmeter connected across the supply reads E volts and an ammeter connected in series with the circuit reads I amps we would conclude that the circuit is consuming $E \times I$ units of power. But the product of the meter readings is only the apparent power; the real power consumption is $EI \cos \theta$. Hence the ratio of real power to apparent power is $\cos \theta$, and is called the *power factor* of the circuit. Apparent power is measured in voltamperes (VA), while real power is measured in watts (W).

Turning back to the impedance triangle in Fig. 104, we see that $\cos \theta = \frac{R}{Z}$, so we may say that the power factor, which is also the cosine of the phase angle, is the ratio of resistance to impedance.

Fig. 111 shows that the apparent power is the vector sum of the real

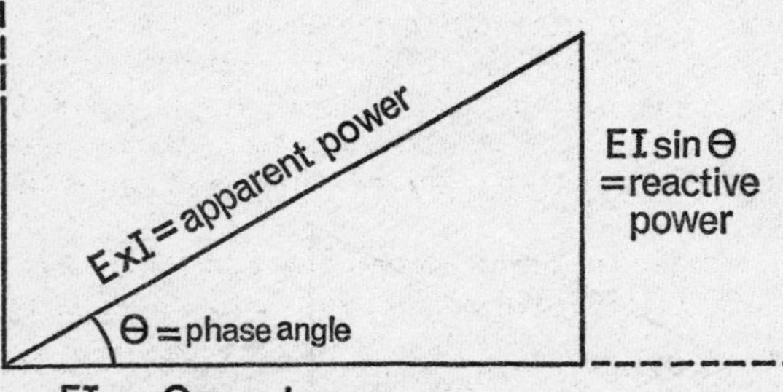

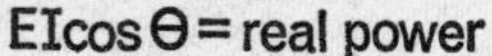

Fig. 111. Power triangle showing relationship between apparent, real, and reactive power

power consumed in a circuit and the *reactive power* alternately stored and returned to the generator by the inductive and capacitive circuit components. In the extreme case when there is no reactive power the phase angle is zero ($\theta = 0$) and the power factor is unity ($\cos \theta = \cos 0 = 1$).

Although the Generating Boards charge their customer only for the real power consumed, they must make provision for the additional reactive power. Transmission cables have to be big enough to carry both the in-phase (resistive) and the out-of-phase (reactive) currents. It is therefore highly desirable to keep the power factor as close to unity as possible.

If the power factor is low owing to highly inductive circuits, such as induction motors, chokes, and transformers (which introduce a lagging phase angle), it can be corrected by a capacitor (connected across the load) whose leading phase angle corrects the inductive lag. What happens is that when the inductor wants storage energy the capacitor is getting rid of it; and when the capacitor wants energy for storage the inductor is just ready to give it. Thus the thirst for stored energy is satisfied *inside* the circuit, and the supply is called upon only to provide for any inequality. The circuit 'looks' to the supply more and more like just a resistance, and the phase angle between the voltage and the current is more nearly reduced to nil. If the two energy storages are exactly equal the circuit does look to the supply exactly like a pure resistance, and the current is exactly in phase with the voltage. This condition we call *resonance*.

RESONANCE

Since the reactance X_L of an inductor increases with frequency, while the reactance X_c of a capacitor decreases with frequency, there is one particular frequency at which they are exactly equal. At this frequency, known as the *resonant frequency*, the net reactance of a circuit containing an inductor and a capacitor is nil ($X_L - X_C = 0$).

If the inductor and capacitor are connected in series the impedance of the circuit reaches its minimum value at the resonant frequency, and the greatest possible current flows; if they are connected in parallel the impedance reaches its maximum value, and the smallest possible current flows.

Inductive reactance X_L is related to frequency by the formula $X_L = 2\pi fL$ (see p. 158) and capacitive reactance X_c is related to frequency by the formula $X_c = \frac{1}{2\pi fC}$ (see p. 157), when L is the inductance and C is the capacitance of a circuit. At the resonant frequency f_r

$$X_L = 2\pi f_r L$$

and

$$X_C = \frac{1}{2\pi f_r C}$$

Since at this frequency $X_L = X_c$, we can put

$$2\pi f_r L = \frac{1}{2\pi f_r C}$$

$$f_r^2 = \frac{1}{4\pi^2 LC}$$

Hence

$$f_r = \frac{1}{2\pi\sqrt{LC}} \tag{4}$$

where f_r is the resonant frequency in hertz or cycles per second, L is the inductance in henrys, and C is the capacitance in farads.

SERIES-RESONANT CIRCUIT

The impedance of a series-connected circuit is

$$Z = \sqrt{R^2 + (X_L - X_C)^2}$$

as shown on p. 169. At resonance $X_L = X_C$, so the impedance becomes

$$Z = \sqrt{R^2 + 0} = R$$

i.e. the total opposition to alternating current is simply the resistance of the circuit, and this is plainly its smallest possible value.

The phase angle θ is given by

$$\tan \theta = \frac{X_L - X_C}{R}$$

Since $X_L = X_C$, $\tan \theta = 0$, and so $\theta = 0$. Hence the current and voltage are *in phase*, and the power factor is unity ($\cos \theta = 1$). In other words, the circuit behaves at the resonant frequency as if it contained resistance alone.

The current in the circuit is given by the ratio of the applied voltage to the impedance. But the impedance at the resonant frequency is equal to the resistance; hence $I = \frac{E}{R}$ and the circuit obeys Ohm's law.

FREQUENCY METER

The circuit shown in Fig. 112 is a form of Wheatstone bridge (p. 32) adapted for alternating current. Three of the arms contain fixed

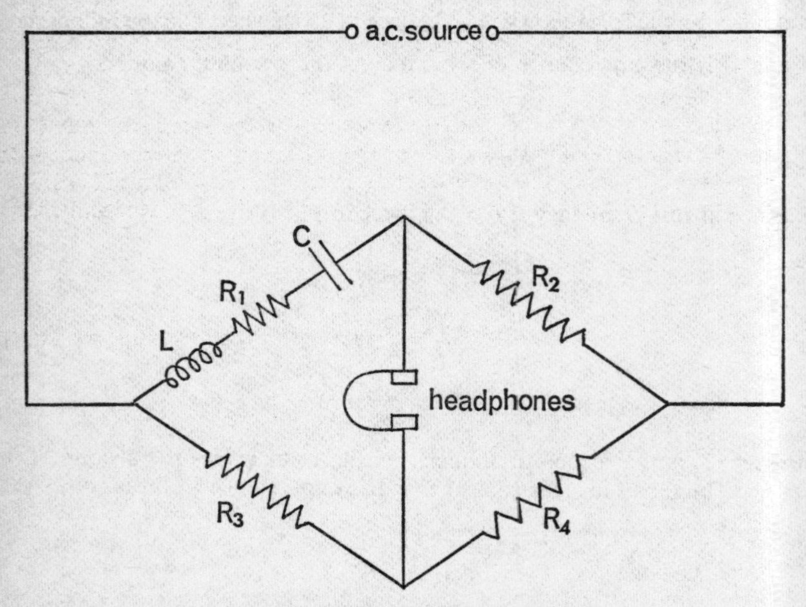

Fig. 112. Frequency-meter circuit

resistors R_2, R_3, and R_4; the fourth arm contains an inductor L and a variable capacitor C, whose combined resistance is represented by R_1. Headphones replace the galvanometer, and until balance is achieved alternating current passing through the headphones produces an audible sound.

The bridge can be balanced only when all four arms contain resistance: reactance (X_L and X_C) cannot be balanced against resistance. Hence the condition for balance is that L and C together act as a resistance, i.e. at resonance when $X_L - X_C = 0$.

The variable capacitor is adjusted until the absence of sound in the headphones indicates that balance has been achieved. We then know that $X_L = X_C$ and the frequency of the a.c. source must be the resonant frequency $f_r = \frac{1}{2\pi\sqrt{LC}}$. If the values of L and C are known the a.c. frequency can be calculated; hence this circuit is known as a frequency meter.

Q FACTOR

The effective (r.m.s.) current flowing in a series circuit is plotted in Fig. 113 against the frequency. The current rises steeply to its maximum value at the resonant frequency f_r, and then falls off rapidly at higher frequencies. The lower curve, drawn for the same circuit except that the resistance has been changed from 7·5 Ω to 12·4 Ω, approaches its maximum more gradually, with its sides (called 'skirts') sloping out over a wide range of frequencies.

The sharpness of the resonance peak is determined by the ratio of the energy stored to the rate of energy loss in the resonant circuit, and can be shown to equal the ratio of the *reactance* of the inductor at the resonant frequency to the resistance of the whole circuit. This ratio is known as the 'quality factor', 'Q factor', or just 'Q' of the circuit. Using symbols

$$Q = \frac{X_L}{R} = \frac{2\pi fL}{R}$$

Since $X_L = X_C$ at resonance, the Q factor could be equally well defined

$$Q = \frac{X_C}{R} = \frac{1}{2\pi fCR}$$

However, since the windings of the inductor account for most of the resistance of a resonant circuit, it is more usual to find Q applied to coils rather than capacitors.

PARALLEL RESONANT CIRCUIT

Resonance also occurs in a circuit containing an inductor and a capacitor connected in parallel, as shown in Fig. 114. At the resonant

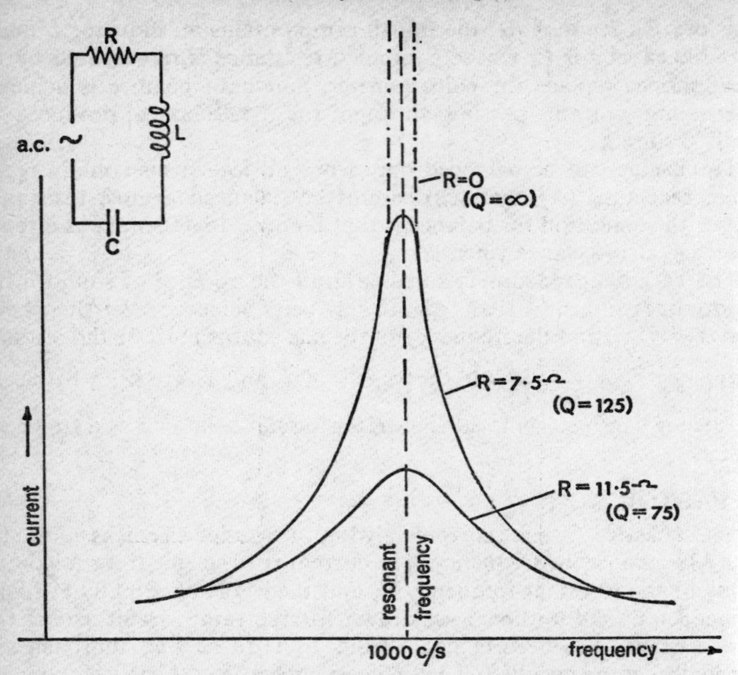

Fig. 113. Resonance curves for a series circuit

frequency f_r the reactance X_C of the capacitor is equal to the reactance X_L of the inductor:

$$X_c = X_L$$

$$\frac{1}{2\pi f_r C} = 2\pi f_r L$$

and

$$f_r = \frac{1}{2\pi\sqrt{LC}}$$

just as it is for a series-resonant circuit. But in this case the impedance is maximum at the resonant frequency, and the current drawn from the source is minimum, as shown in Fig. 115.

The total impedance at the resonant frequency can be shown to be

$$Z = \frac{X_C X_L}{R}$$

$$= \frac{1}{R} \times \frac{1}{2\pi f_r C} \times 2\pi f_r L$$

$$= \frac{L}{CR}$$

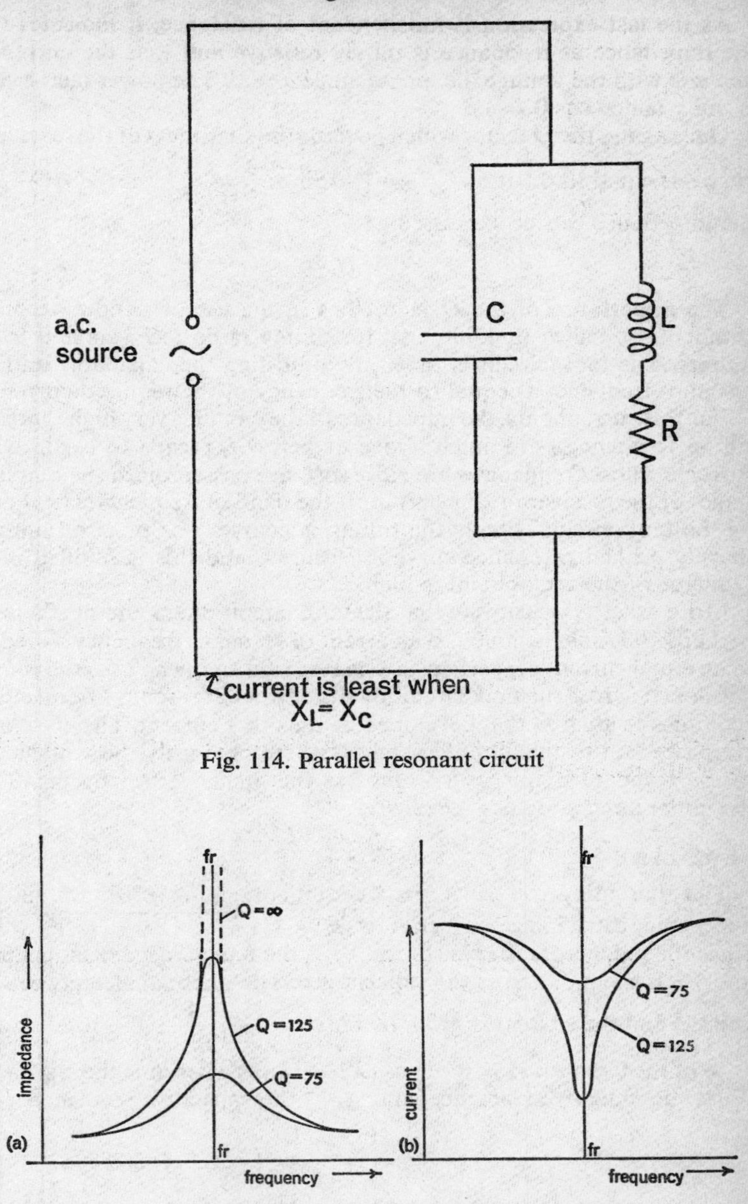

Fig. 114. Parallel resonant circuit

Fig. 115. Resonance curves for a parallel circuit: (*a*) impedance; (*b*) current

As the last expression is independent of reactance, it indicates that the impedance at resonance is purely resistive and that the current is *in phase* with the voltage, i.e. phase angle $\theta = 0$. The power factor $\cos \theta$ is unity (since $\cos 0 = 1$).

Once again, the Q factor which governs the sharpness of the resonance curves is equal to the ratio $\frac{X_L}{R} = \frac{X_C}{R}$. Since $Q = \frac{X_C}{R}$ and $Z = \frac{X_C X_L}{R}$, the impedance can be expressed as

$$Z = QX_L$$

The importance of the Q factor lies in the use of parallel resonant circuits (also called tank circuits) for tuning radio receivers and transmitters. The tank circuit is tuned, by adjusting the capacitor, until its resonant frequency is equal to the frequency of the required current.

For this current only, the impedance of the circuit is very high: currents whose frequencies are much above or below f_r escape to earth, while currents whose frequencies are close to f_r are passed on to the amplifier stages of the receiver or transmitter. If the band of frequencies held back by the tank circuit is wide the tuning is coarse. For precise tuning, a sharply peaked resonance curve is required, and this is fulfilled if the Q factor of the tank circuit is high.

More exactly, we can say that the tank circuit offers an impedance of perhaps 100 times as much to a current of resonant frequency as it does to an equal current of, perhaps, twice resonant frequency. So the *voltage* developed across the tank circuit by the current at resonant frequency is 100 times as great as that developed by the other current. These voltages are passed on to the amplifier stages, where obviously the voltage due to the resonant-frequency current has the greater effect by far. Thus the tank circuit provides *selectivity*.

SUMMARY

The alternating voltage across a circuit containing resistance, inductance, and capacitance in *series* is $E = \sqrt{V_R^2 + (V_L - V_C)^2}$, where V_R is the p.d. across the resistance, V_L is the p.d. across the inductance, and V_C is the p.d. across the capacitance. The phase angle between the voltage and the current is given by $\tan \theta = \frac{V_L - V_C}{V_R}$. The total impedance of the circuit is $Z = \sqrt{R^2 + (X_L - X_C)^2}$, when R is the resistance, X_L is the inductive reactance, and X_C is the capacitive reactance.

$$\frac{X_L - X_C}{R} = \tan \theta$$

For resistance, inductance, and capacitance connected in *parallel*, the total current is $I_t = \sqrt{I_R^2 + (I_L - I_C)^2}$, when I_R is the current in the resistance, I_L is the current in the inductance, and I_C is the current in the capacitance. The phase angle between current and applied voltage is given by $\tan \theta = \frac{I_L - I_C}{R}$.

The *real* power in an a.c. circuit is $P = EI \cos \theta$ (watts), where E is the applied voltage, I is the current, and θ is the phase angle. Cos θ is the *power factor*. $E \times I$ is the *apparent power* (measured in voltamperes). Apparent power is the vector sum of real power and *reactive power*.

Resonance occurs when the inductive reactance X_L of a circuit is equal to the capacitive reactance X_C. The frequency at which this happens is the *resonant frequency* f_r

$$f_r = \frac{1}{2\pi\sqrt{LC}}$$

For a series circuit at resonance, current is maximum, the impedance is equal to the resistance alone ($Z = R$); the phase angle is 0 and the power factor is unity.

For a parallel circuit at resonance, current is minimum, the impedance is $Z = \frac{R}{X_C X_L} = \frac{L}{CR}$ and is maximum, the phase angle is 0 and the power factor is unity.

The sharpness of a resonance curve is governed by the ratio $\frac{X_L}{R}$, called the Q factor.

CHAPTER ELEVEN

A.C. GENERATORS AND MOTORS

We saw in Chapter Seven how a coil rotating in a magnetic field generates an *alternating* e.m.f. of the type which is supplied commercially through the domestic a.c. mains. The machines which generate alternating e.m.f.s in power stations are called *alternators* (the term 'dynamo' is confined to d.c. generators), and for practical reasons differ considerably in construction from the simple rotating coil. In order to connect alternating e.m.f. induced in a rotating coil to the external

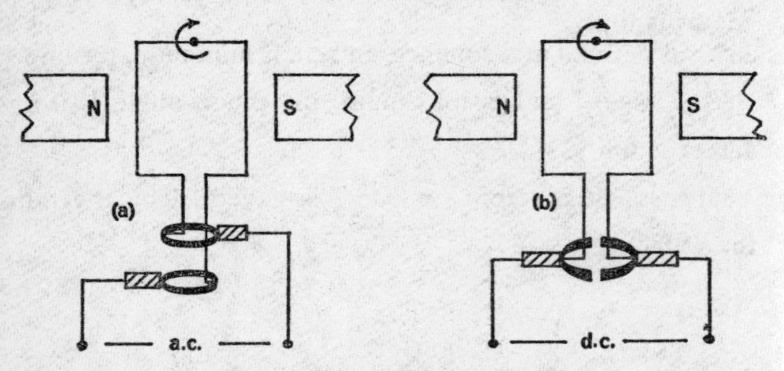

Fig. 116. (*a*) Slip rings on a.c. generator. (*b*) Commutator on d.c. generator

circuit, each end of the coil is attached to a 'slip ring' while the external circuit is connected to carbon brushes as shown in Fig. 116 (*a*). (If each end of the coil is attached to a segment of a commutator, as shown in Fig. 116 (*b*), the external circuit connected to the brushes carries a pulsating form of direct current.)

In a generator developing, say, 11,000 volts between its output terminals, the problems of insulation and mechanical construction associated with rotating slip rings and stationary brushes are tremendous and it is not surprising that the design of practical generators eliminates the need for slip rings. This is achieved by keeping the coil stationary and rotating the magnetic field. Fig. 117 (*a*) shows a magnet rotating between a pair of fixed coils; the magnetic field cutting each coil rises and falls as the poles move past it and an alternating e.m.f. is induced. Since the N pole is passing one coil while the S pole is passing the other, the two induced e.m.f.s are out of phase by 180° (i.e. the positive peak of one coincides with the negative peak of the other). The two coils are therefore connected in 'series opposition' (see p. 151) so that the alternating e.m.f. at the terminals is the sum of the voltages induced in each coil.

In practice, the rotating permanent magnet is replaced by an electromagnet, as shown in Fig. 117 (*b*). Since there are now two sets of windings, we have to distinguish between them by calling the rotating electromagnet the *rotor* and calling the stationary armature the *stator*.

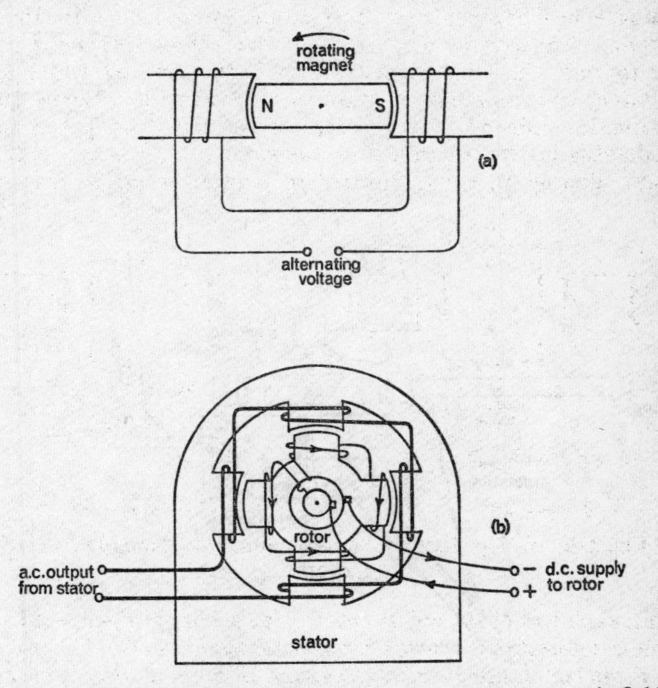

Fig. 117. Generating an e.m.f. by means of a rotating magnetic field: (*a*) permanent magnet; (*b*) electromagnet

The windings of the rotor, which provide the magnetic field, are supplied with direct current obtained from a small separate d.c. generator called the *exciter*. The exciter is usually mounted on the same shaft as the main alternator and driven from the same power source (usually a steam turbine). Current from the exciter enters the rotor via carbon brushes and slip rings, but since the voltage is low, the insulation problems mentioned above are not serious.

The core of the rotor is a nickel-steel forging with slots cut along its length to accommodate the windings. A certain amount of heat is produced in the rotor by eddy currents, by hysteresis, and by the resistance of the windings; so it has to be ventilated by forcing air or hydrogen over its surface and through channels cut in the core. Hydrogen is chosen because it has a high specific heat and low viscosity, i.e. it does not greatly resist the rotation of the rotor.

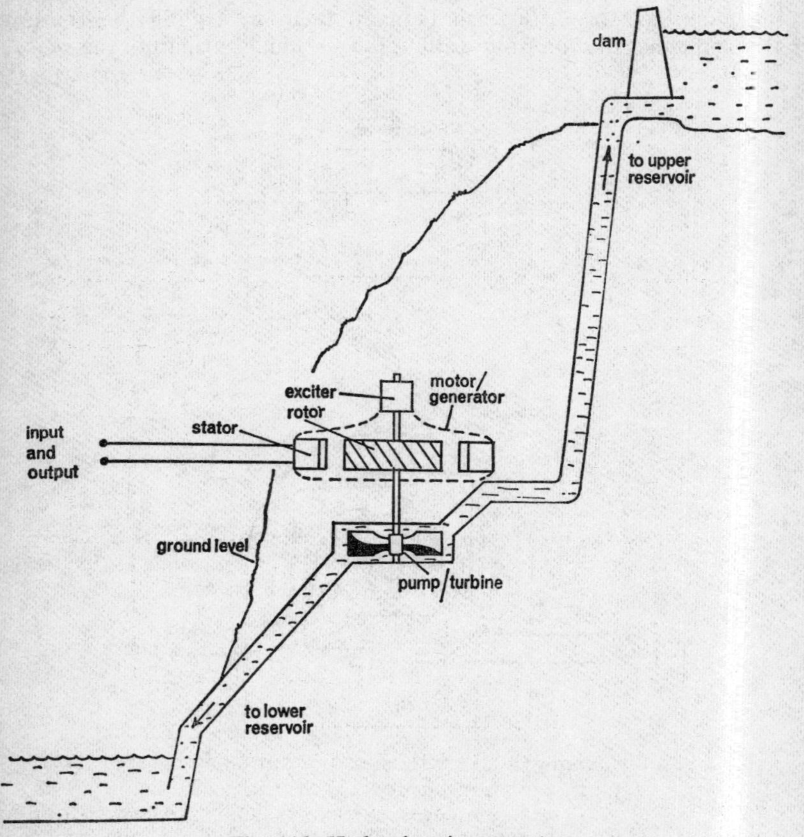

Fig. 118. Hydroelectric generator

The stator is built up from thin 'soft' steel laminations to reduce eddy-current losses. Like the rotor, slots are cut in the stator to carry copper windings. Heat is removed by water circulating in tubes passing through the laminations. The gap between the rotor and the stator is usually made much wider in a generator than it is in a motor. For various technological reasons it is desirable to have some flux leakage in an a.c. generator, whereas in an a.c. induction motor (see p. 190) it is desirable to reduce leakage to a minimum, since the magnetic field is produced by an alternating current and we must avoid excessive magnetic storage energy surging into and out of the machine and giving it a low power factor (see p. 176).

An alternator having two poles (Fig. 117 (*a*)) generates one cycle of e.m.f. per revolution; to generate 50 Hz it must rotate at 50 rev/sec or 3,000 rev/min. A generator having four poles (Fig. 117 (*b*)) generates 50 Hz when it rotates at 1,500 rev/min.

In a generating station the alternators are connected together in

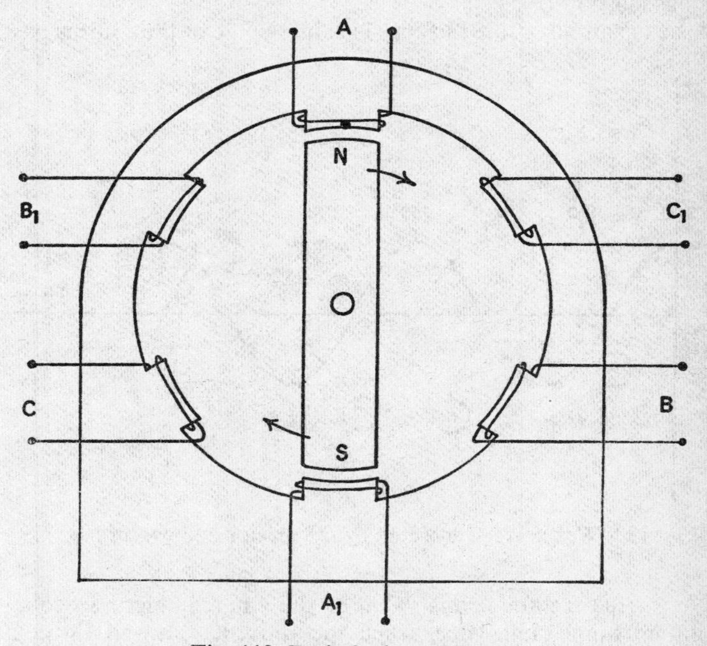

Fig. 119. Basic 3-phase generator

parallel. As the demand increases, extra alternators are switched in, but only when they have been exactly synchronized with those already working. Hence the speed of rotation must be carefully controlled. Alternators driven by steam turbines rotate at high speeds and are mounted horizontally. In hydroelectric generating stations the alternators are driven by water tubines and rotate relatively slowly; they are mounted on a vertical axis, as shown in Fig. 118.

The generator illustrated in Fig. 118 is rather special. Normally, the water in the upper reservoir rushes downhill and forces the turbine to rotate, turning the alternator, which is mounted on the same shaft.

When the demand for electricity is low, however, the whole process is put into reverse. Power is then drawn from the mains to supply the alternator, which becomes a motor; the turbine blades, turned by this motor, act as a pump lifting water back into the upper reservoir, where it is stored until needed again for generating power.

In actual practice, power is generated commercially in the form of a 3-phase current rather than the single-phase current of the domestic supply. The simple single-phase alternator shown in Fig. 117 is modified to produce a 3-phase output by giving it three separate sets of stator windings (Fig. 119).

At the instant illustrated in Fig. 119 the N pole of the rotor is passing

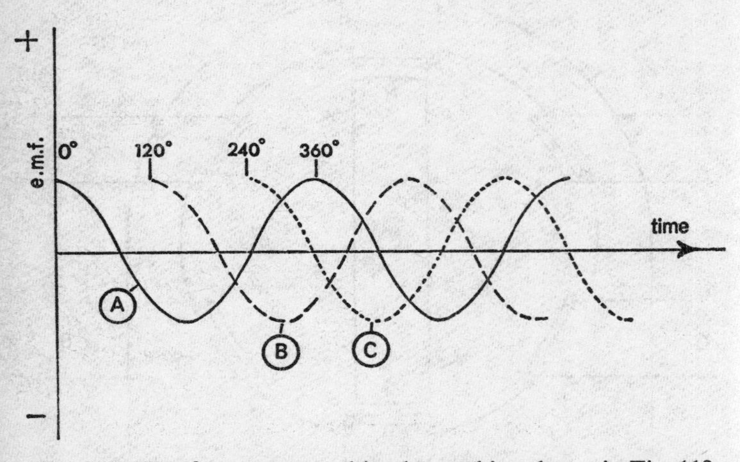

Fig. 120. Waveforms generated by the machine shown in Fig. 119

coil A, so the instantaneous value of the e.m.f. induced in coil A is maximum. Some time later, when the rotor has moved through an angle of 120°, the N pole is passing coil B and the induced e.m.f. therein is maximum. Meanwhile the e.m.f. in coil A has changed polarity and is approaching its minimum value, which it will reach when the S pole of the rotor faces it. Later still, when the rotor has turned through a further 120°, the e.m.f. in coil C is maximum, while the e.m.f. in coil A has begun to increase again and is approaching zero. These three induced e.m.f.s are shown in Fig. 120. It will be seen that they are out of phase with one another by 120°.

The rotor of a practical 3-phase generator is, of course, an electromagnet supplied with d.c. from an exciter; it has several pairs of poles arranged symmetrically around its circumference, while the stator has three times as many pairs of windings.

A.C. MOTORS

Large commercial generators cost something in the region of £1,000,000, so naturally they must be employed as economically as possible. Because 3-phase generation exploits these machines much more effectively than single-phase generation, this is the form in which power is in fact produced. Another reason for generating and transmitting power as a 3-phase current is that it has considerable advantages over single-phase current for driving electric motors.

In construction, electric motors are almost identical with generators. In fact, as we saw in Chapter Six, a d.c. motor actually generates a

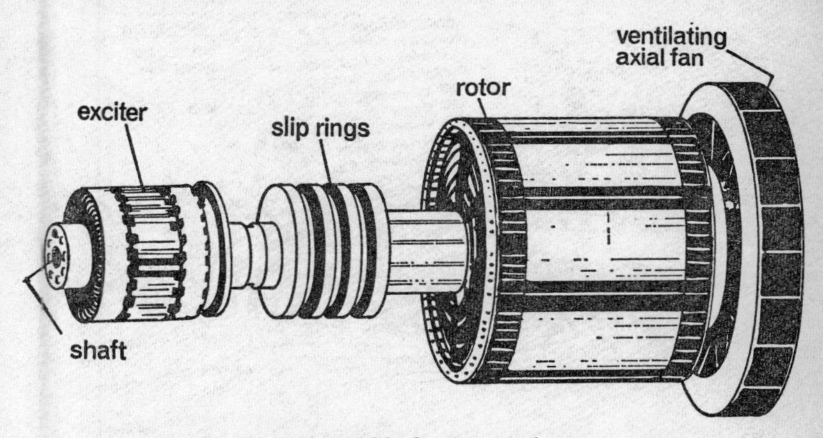

Fig. 121. Rotor assembly from a synchronous motor

back e.m.f. all the time it is running. Similarly, the 3-phase alternator described above can be used as a motor if power is supplied to it instead of taken from it. The 3-phase mains current is fed into the three sets of stator windings, while the rotor is supplied with direct current from the exciter. We then have what is known as a *synchronous motor*, the name being derived from the fact that, once the motor has run up to full speed, it runs synchronously with the alternators supplying it with power.

The synchronous motor runs at an absolutely constant speed, provided that the mains frequency does not alter. Although it is not possible to vary the speed, a choice of two or three different speeds can be obtained by having the poles interconnected in two or three different ways. The machine shown in Fig. 121 is a two-speed motor, and therefore has two pairs of slip rings.

Of all the various types of electric motor, easily the most popular is the 3-phase 'squirrel-cage' *induction motor*. It is very simple in construction, very reliable, and very robust. The name 'squirrel-cage' comes from the appearance of the rotor, which is a cylindrical cage of copper bars firmly brazed to circular end rings or integrally cast with them. The cage is completely filled with 'soft' steel laminations.

The stator is also built up from steel laminations, clamped together, and having rectangular slots to carry the windings of insulated copper wire; it is very similar to the stator of an alternator.

There is no electrical connexion to the rotor, and hence no slip rings. The alternating current in the stator windings sets up an alternating magnetic field which is cut by the bars of the rotor. We can think of

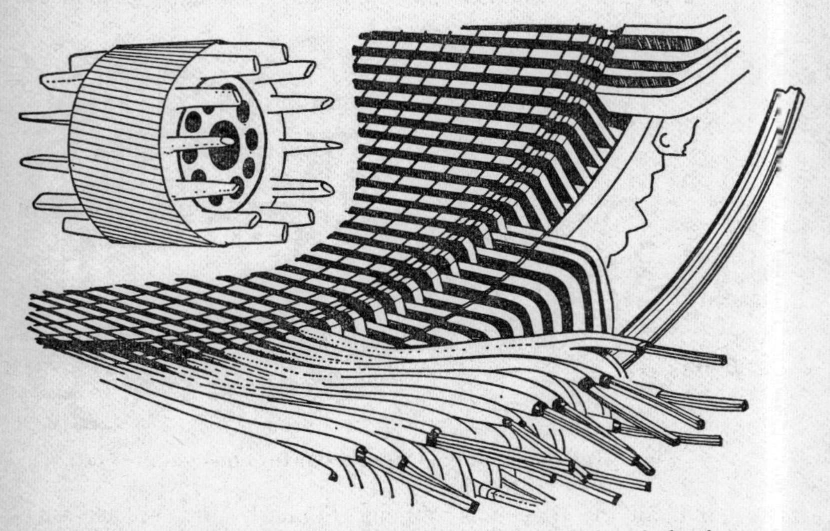

Fig. 122. Rotor assembly and stator construction of a 3-phase induction motor

this motor as being a transformer in which the stator windings are the primary and the rotor bars are the secondary. The secondary has a very low resistance, since it is short-circuited by the end rings, and thus a very large current is *induced* in the rotor bars: hence the name 'induction motor'. Small- and medium-sized induction motors have aluminium rotors, in which the bars, end rings, and cooling fins are cast from high-purity aluminium. As shown in Fig. 122, the bars are skewed so that they do not lie exactly parallel to the stator windings; this improves the starting torque and reduces vibration when running. The rotor is shrunk on to a carbon-steel shaft which may also carry an axial fan.

The principle of the 3-phase induction motor is remarkably simple. Each phase of the current is fed to each of three systems of windings on the stator, as shown in Fig. 123. As the three currents A, B, and C grow and decay in sequence, so the magnetic field set up in the stator *rotates* in the direction A ⟶ B ⟶ C. Alternating currents are induced in the bars of the rotor, and the interaction between the magnetic field of the induced current and the rotating field causes the squirrel cage to rotate.

When the motor is driving a load the rotor lags slightly behind the rotating field; the difference between the speed of the rotor and the speed of the rotating field is called 'slip', usually expressed as a per-

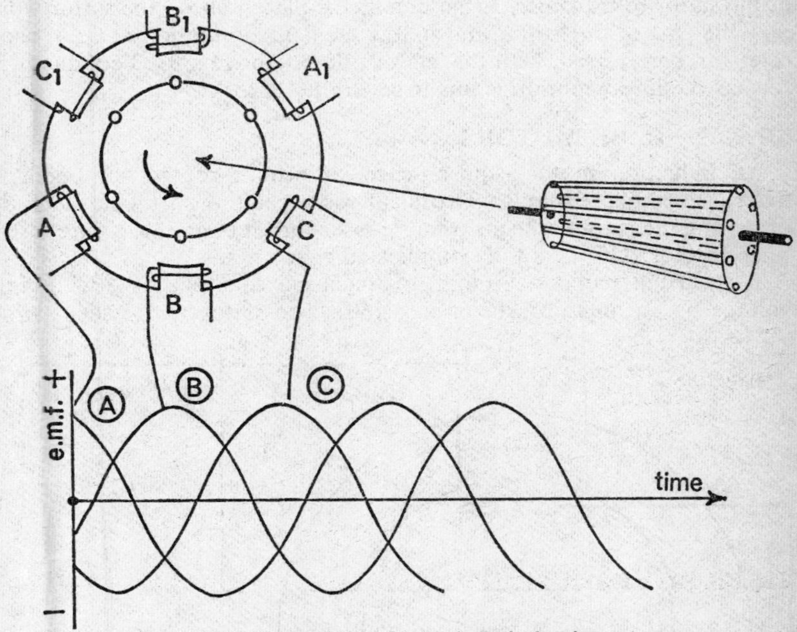

Fig. 123. Rotating field of a 3-phase induction motor

centage of the rotating-field speed. The speed cannot be varied at will—this is the one big drawback with squirrel-cage motors—but it can be reversed by transposing the connexions of *one* of the three phases.

The speed can be varied to some extent if the squirrel cage is replaced by windings like those of the stator and brought out to slip rings and carbon brushes. The slip-ring winding is like that on the rotor of an a.c. generator, with insulated wire in a slotted core. The brushes collect the current induced in the rotor and dissipate it in a variable resistor. As the resistance is reduced, the rotor speed increases, and vice versa.

LINEAR MOTOR

A great many of the uses to which electric motors are put, e.g. cranes, conveyor belts, loom shuttles, and, most important, railway traction, really call for a straight-line motion rather than the turning motion of a rotor. For these purposes a *linear motor* has been developed. Basically the linear motor may be regarded as a 3-phase induction motor whose stator has been opened along its length and flattened out. What was the rotating field is now a linear field which travels from one end of the stator to the other. A flat conductor placed above the stator will play the part of the 'rotor' and follow the field along the stator; or the roles can be reversed, with the 'stator' placed above a fixed conductor, so that the linear motion is imparted to the 'stator'.

SINGLE-PHASE MOTORS

An induction motor, squirrel-cage or wound-rotor, will operate from a single-phase supply: the difficulty is to get it to start. One method of getting the rotor to begin rotating is to turn it temporarily into a 2-phase motor. This is not as complicated as it sounds.

The current passing through a capacitor is out of phase with the voltage by an angle of 90° (see p. 160). The stator is provided with

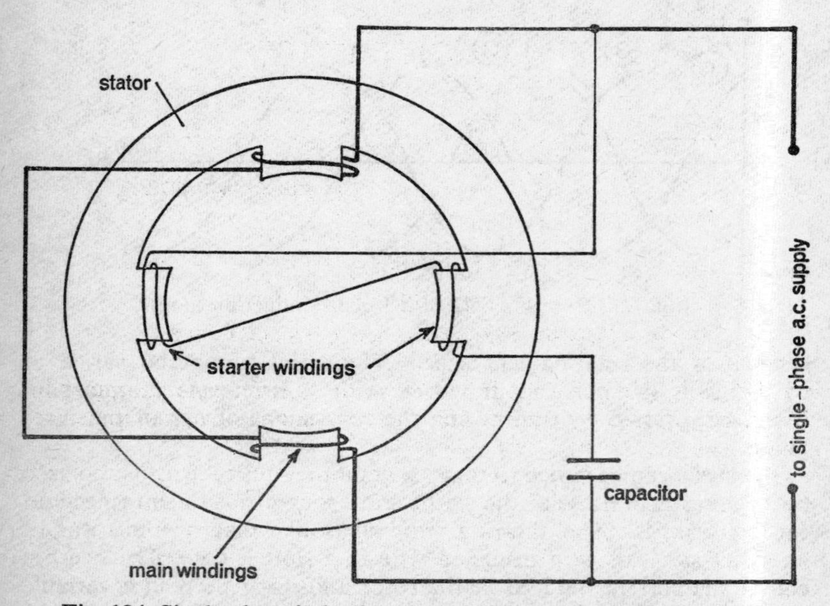

Fig. 124. Single-phase induction motor with starting capacitor (rotor not shown)

extra windings at 90° to the main windings, and these are supplied through the capacitor. So the currents in the two windings, differing in phase by 90°, set up a rotating field, which the rotor follows. When the rotor has reached its synchronous speed it will continue to rotate without the assistance of the starting winding, which may then be cut out. The direction of rotation can be reversed by exchanging the connexions of the main windings or of the starting windings, but not both.

A single-phase induction motor of simpler construction is shown in

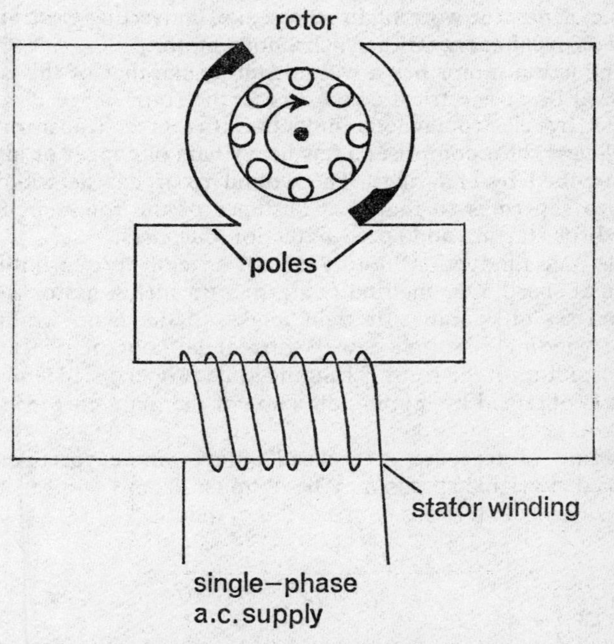

Fig. 125. Schematic shaded-pole induction motor

Fig. 125; this is called a shaded-pole motor. It is extremely popular in clocks, gramophones, tape recorders, and similar applications requiring a small constant-speed drive.

A *universal* motor is one which can be operated from either a.c. or d.c. supplies. As explained on p. 102, a d.c. motor will operate with alternating current, and it has the big advantage that the speed can be varied by altering the current with a rheostat. The universal motor is therefore preferable to an induction motor in applications where variable speeds are required. It is basically the same as the series-connected d.c. motor described on p. 100. Both rotor and stator are laminated, and carry insulated copper windings, the ends of the rotor winding being brought out to commutator segments making contact with carbon brushes.

SUMMARY

In an a.c. generator, called an alternator, e.m.f. is induced in a series of fixed windings called the stator, by a magnetic field produced in a series of rotating coils called the rotor. The rotor is energized with direct current from a d.c. generator, called the exciter, mounted on the same shaft as the rotor. Three-phase current is obtained from three separate systems of stator windings, displaced at 120° to one another.

An a.c. generator working in reverse, i.e. converting electrical energy into mechanical energy, is a synchronous motor.

An induction motor has a wound stator, like that of the alternator; there need be no electrical connexion to the rotor, since its current is obtained by electromagnetic induction (as in a transformer). The squirrel-cage rotor comprises a few heavy bars of copper or aluminium, short-circuited by end rings. The wound rotor has its windings connected to slip rings so that the resistance of the rotor circuit can be adjusted for starting and speed-reduction purposes.

Single-phase induction motors do not develop torque until they are running at speed. One method of starting up such a motor is to give it a second set of windings at right angles to the main windings. The starting winding is supplied with current 90° out of phase with the main current until the rotor is running at its proper speed; out-of-phase current is obtained by tapping off some of the main current through a capacitor.

Induction motors are at a disadvantage where variable speed is concerned. Variable speeds may be obtained from a 'universal' motor, which is basically a series-connected d.c. motor.

CHAPTER TWELVE

OTHER METHODS OF GENERATING E.M.F.

The e.m.f. in a circuit is nearly always provided either by a battery, as described in Chapter Five, or some kind of rotating generator as described in Chapter Eleven. However, there are several other methods of obtaining an e.m.f. which are of great technical interest if not of such commercial importance as the battery and dynamo.

M.H.D.

Magnetohydrodynamic (m.h.d.) generation is, in principle, a very attractive method of obtaining electrical energy from a fuel without the

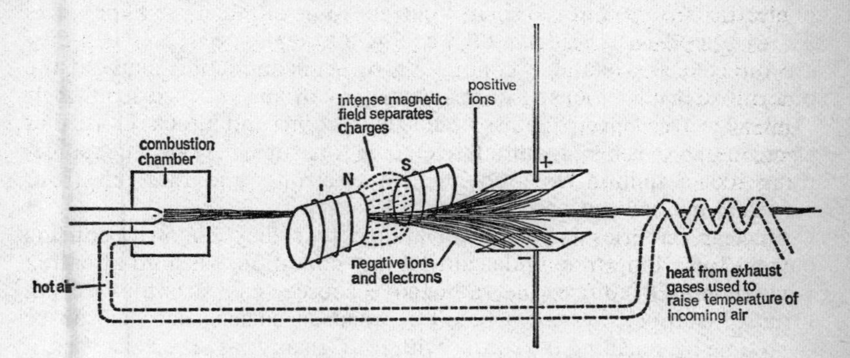

Fig. 126. Principle of m.h.d. generation

need for rotating machinery. The products of burning a fuel such as coal or oil in air are energy in the form of heat, gases, and perhaps solid ash. While the gas molecules are at a very high temperature, their energy of vibration is so great that they tend to throw off electrons and become positively charged ions. Thus the hot gas mixture contains a 'plasma' of free electrons and positive ions.

If, as shown in Fig. 126, this mixture is blown through a magnetic field the ions will be deflected upward and the electrons downward (or vice versa). Hence a conductor placed above the field continually receives a positive charge, while a conductor placed below the field receives a negative charge. The two conductors can then be used as a source of e.m.f., just like the terminals of a battery. The plasma is a *conducting* gas, and the arrangement is therefore a way of moving a conductor rapidly in a magnetic field: essentially the same thing that happens in a conventional d.c. generator.

Although m.h.d. is theoretically a very efficient method of generating electrical energy, it presents practical difficulties which have yet to be completely solved.

NUCLEAR BATTERY

Another very recent development, though on a much smaller scale, is the so-called nuclear battery. As is well known, the nuclei of radioactive atoms are unstable and tend to throw out α-, β-, and γ-rays. Now β-rays are streams of fast-moving electrons, which may seem strange, since in the picture of an atom with which we began Chapter One the nucleus contains positive protons and uncharged neutrons, but no electrons. In fact, each of these electrons (which we will call a β-particle, to avoid confusion with the orbital electrons lying well outside the nucleus) are generally agreed to come from the transformation of a neutron into a proton plus a β-particle. Since β-particles are electrons, a β-ray is a form of electric current, but too small a current to be of much use in practice.

However, β-particles have high energy and high speed; so when they are directed at a suitable conducting or semiconducting material the outermost electrons of the atoms in their path are knocked from their orbits. The freed electrons also collide with atoms and knock out further electrons, so the effect is cumulative. Each β-particle may produce something like a million slow-moving free electrons, and these electrons represent a small but useful current of electricity.

Nuclear batteries have the advantage that they go on producing current for a long time under almost any conditions with no attention whatsoever. Of course, the radioactive source (e.g. strontium-90, an artificial isotope) presents a possible radiation hazard, but this can be overcome by shielding the source with lead. In any case the chief uses of nuclear batteries to date have been in Earth satellites and marine-navigation buoys, where any leakage of β-rays would be harmless.

Nuclear batteries should not be confused with *nuclear reactors*. The reactor in a nuclear power station produces *heat* by controlled fission (splitting) of uranium-235 or plutonium-239. The energy released when one atom splits is almost insignificant, but vast quantities of heat are produced by a *chain reaction*. This heat is removed by a liquid or gas circulating through pipes built into the reactor, and taken to a 'heat exchanger', where, in present-day nuclear power stations, it is used to generate steam (Fig. 127). Steam raised in the heat exchanger drives a turbine, which in turn drives a conventional alternator as described on p. 185.

Thus a nuclear power station employs just as many steps in the production of electricity as a coal- or oil-fired station, and each step inevitably wastes some energy. But of course once the reactor has been started up there is no need to feed it with a constant stream of fuel, since 1 lb of uranium is equivalent to at least 1,000 tons of coal. Obviously the efficiency would be greatly enhanced if the heat energy in the reactor

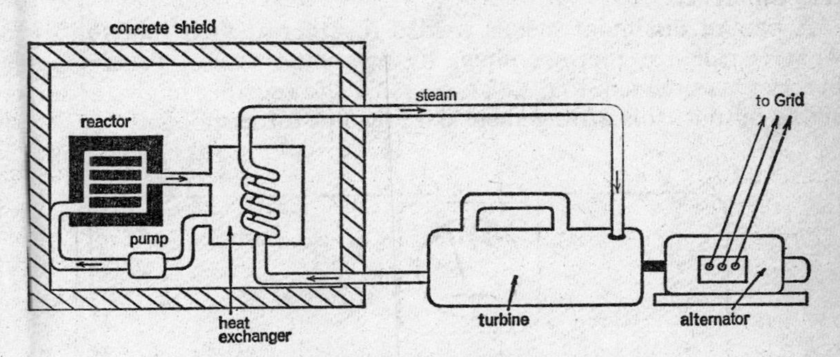

Fig. 127. Schematic of a nuclear power station

could be converted to electrical energy directly; perhaps by the m.h.d. principle described above, or by using the *thermoelectric effect*.

THERMOELECTRIC EFFECT

When two dissimilar metals such as iron and copper are joined at both ends to form a closed circuit, and one of the junctions is at a higher temperature than the other (Fig. 128), a current is set up. The e.m.f.

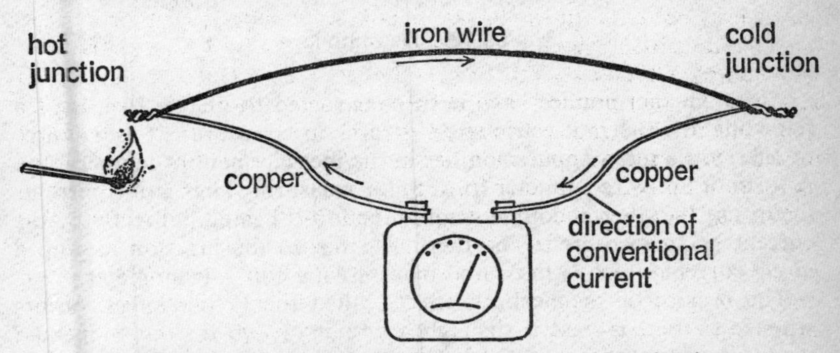

Fig. 128. Thermoelectric effect in a copper–iron circuit

driving this current is called a 'thermoelectric e.m.f.', and the phenomenon is known as the thermoelectric effect or Seebeck effect after the German physicist who discovered it in 1821.

Usually a thermoelectric e.m.f. is very small, only a few millionths of a volt. For a copper–iron circuit it is found to be about 7 μV for every degree Centigrade of temperature difference between the junctions; for antimony and bismuth it is as high as 100 μV per deg C, while for copper and Constantan (55 per cent copper, 45 per cent nickel), the two metals most often used in practice, it is 40 μV per deg C of temperature difference.

A pair of dissimilar metals welded together at their junction forms what is called a thermocouple. By arranging several thermocouples in series, as shown in Fig. 129, the e.m.f.s add together to give an appreciable output; this arrangement is known as a thermopile.

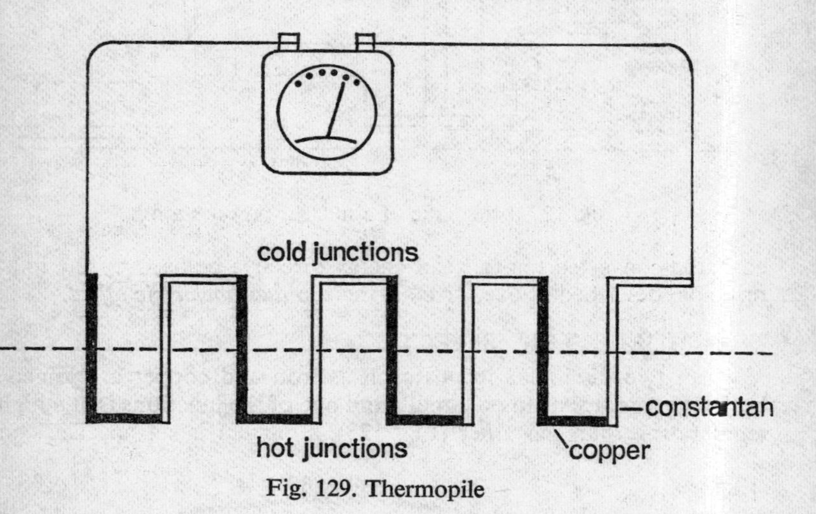

Fig. 129. Thermopile

Although thermopiles have been constructed to deliver e.m.f.s of a few volts, the thermoelectric effect is rarely used at present as a source of energy. Its main application lies in the measurement of temperature. A form of hot-wire ammeter (p. 135) for measuring alternating currents incorporates a thermocouple whose junction is heated indirectly by the current being measured. The thermal e.m.f. in the junction sets up a *direct* current which is measured by a moving-coil galvanometer.

The reverse of the Seebeck effect, called the Peltier effect, occurs when a current is passed through a circuit of two dissimilar metals: heat is produced at one of the junctions and absorbed at the other. The latter junction is therefore cooled, and this cooling effect has been used as the basis of a novel type of refrigerator.

For a simple explanation of the thermoelectric effect we must accept that when two dissimilar metals are pressed together free electrons drift haphazardly across the junction. Because of the different atomic structure of each metal, electrons pass more readily across the boundary in one direction than in the other. This results in a displacement of charges, making one metal positive and leaving the other negative.

The voltage between the metals is called a contact potential difference, and is influenced by the temperature of the boundary. By keeping one junction at a higher temperature than the other, the unequal drift of electrons past each junction maintains a steady potential difference, and a thermoelectric e.m.f. can be obtained.

PHOTOELECTRIC EFFECT

In the same way that the variation of contact potential difference with temperature gives rise to the thermoelectric effect, so the variation of contact potential difference with illumination gives rise to a *photoelectric* effect.

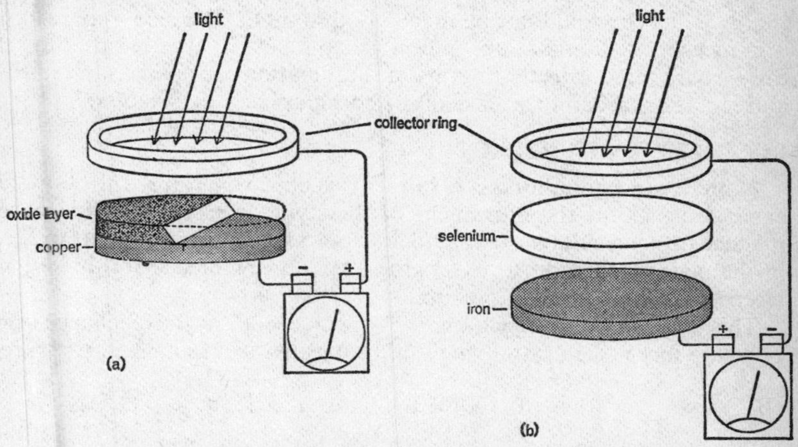

Fig. 130. Photovoltaic cells: (*a*) copper oxide; (*b*) selenium

The 'solar cells' which Earth satellites use for recharging their batteries convert the energy of visible light and ultraviolet radiation into electrical energy directly. They are properly called photovoltaic cells or *barrier-layer* cells.

When light falls on certain semiconductors, such as selenium or cuprous oxide, electrons are stripped from their atoms—how many depending on the intensity of the illumination. The cell shown in Fig. 130 (*a*) consists of a copper disc whose front surface has been oxidized to form a layer of cuprous oxide. The interface between the copper and the oxide acts as a barrier, or rectifier, which allows electrons to pass from oxide to copper but not in the reverse direction. Hence the electrons released by light falling on the front of the cell escape from the oxide, leaving it with a deficit of electrons, and hence a positive charge, while the copper gains a surplus of electrons, and hence a negative charge.

In this way an e.m.f. is generated across the cell, and will drive current through a circuit connected between the copper and the cuprous oxide. The front surface carries a conducting layer of gold, thin enough for light to pass through, so that the external circuit can be connected to the oxide.

Fig. 130 (*b*) shows the construction of the widely used selenium cell. It is a sandwich of three layers: the transparent gold window, the light-sensitive layer of selenium alloy, and an iron base. The action is similar to that of the copper oxide cell, the rectifying barrier being the boundary between the selenium and the iron. An e.m.f. of about one-quarter of a volt is produced in this type of cell; higher voltages can be obtained by connecting cells in series.

One very obvious difficulty of using the photocell as a source of energy is that it would not be much good at night. Its main application is in measuring illumination, since a galvanometer connected to the cell can be calibrated directly in units of illumination, or aperture numbers and shutter speeds in the case of a photographer's light-meter.

PIEZOELECTRIC EFFECT

A modern gramophone is a form of electric generator. In the more expensive pickups the movement of the stylus is converted through a magnet moving relative to a coil into an alternating e.m.f. However, in the majority of record-players the e.m.f. is obtained from the *piezo-electric* effect in a crystal pickup.

This effect is the appearance of positive and negative charges on opposite faces of a crystal when it is subjected to mechanical pressure.

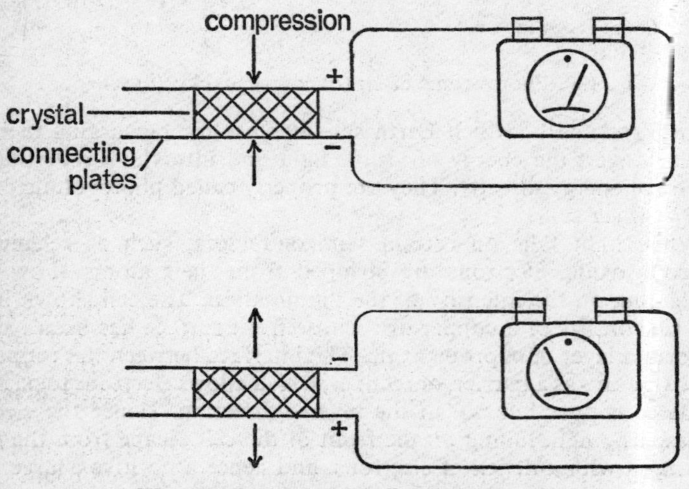

Fig. 131. Piezoelectric effect in a crystal

It occurs in natural crystals, such as quartz, Rochelle salt, and tourmaline, as well as in ceramic materials, such as barium titanate. As shown in Fig. 131, the charges which appear when a crystal is compressed are reversed when it is stretched. The charges exist only as long as the mechanical force is exerted; but a vibrating force, i.e. a succession of compressions and tensions, will generate a continuous alternating e.m.f.

Piezoelectric crystals used in gramophone pickups generate an e.m.f. of no more than 1 V, but certain ceramics are available which generate thousands of volts. One such material (lead zirconate titanate) is used in an automatic ignition system for gas fires: the crystal is squeezed in a kind of nutcracker, and the resulting high voltage produces an electric spark to light the gas.

The piezoelectric effect, like the thermoelectric effect, is reversible. If an alternating e.m.f. is applied to metal plates on opposite faces of a quartz crystal continuous lengthwise vibrations are set up. This effect is used in crystal-controlled radio oscillators and in ultrasonics.

ELECTROSTATIC GENERATORS

A number of laboratory applications, including the production of high-energy X-rays and the acceleration of particles used in bombarding atomic nuclei, call for very high *direct* voltages. An alternating p.d. of perhaps 1,000,000 V could be obtained from a step-up transformer, but it would then need to be rectified. The alternative is to employ an *electrostatic* method, such as the Van de Graaff generator illustrated in Fig. 132.

This generator is remarkably simple in principle and in construction. It consists of a large, smooth aluminium sphere mounted on top of a hollow glass column. An endless belt, made of insulating material such as rubber, runs between pulleys at each end of the column; it is driven by a small electric motor in the base of the apparatus.

The essential action of the Van de Graaff generator is that charge sprayed on to the moving belt finds its way to the outer surface of the sphere, where it gradually accumulates.

The charge comes originally from a high-voltage battery or other d.c. source, one terminal of which is connected to a metal comb mounted close to the moving belt. In Fig. 132 this comb is shown as an arrowhead connected to the negative battery terminal: it therefore acquires a negative charge or electron surplus. Since electrons on the surface of a conductor tend to accumulate around sharp points, the teeth of the comb become strongly charged—so strongly in this case that the air molecules in contact with them are turned into negative ions and carry electrons over to the nearby belt.

Inside the sphere a second comb removes the surplus electrons from the belt in the same way that a pointed lightning-conductor discharges a thundercloud. What happens is that the negative charge on the belt

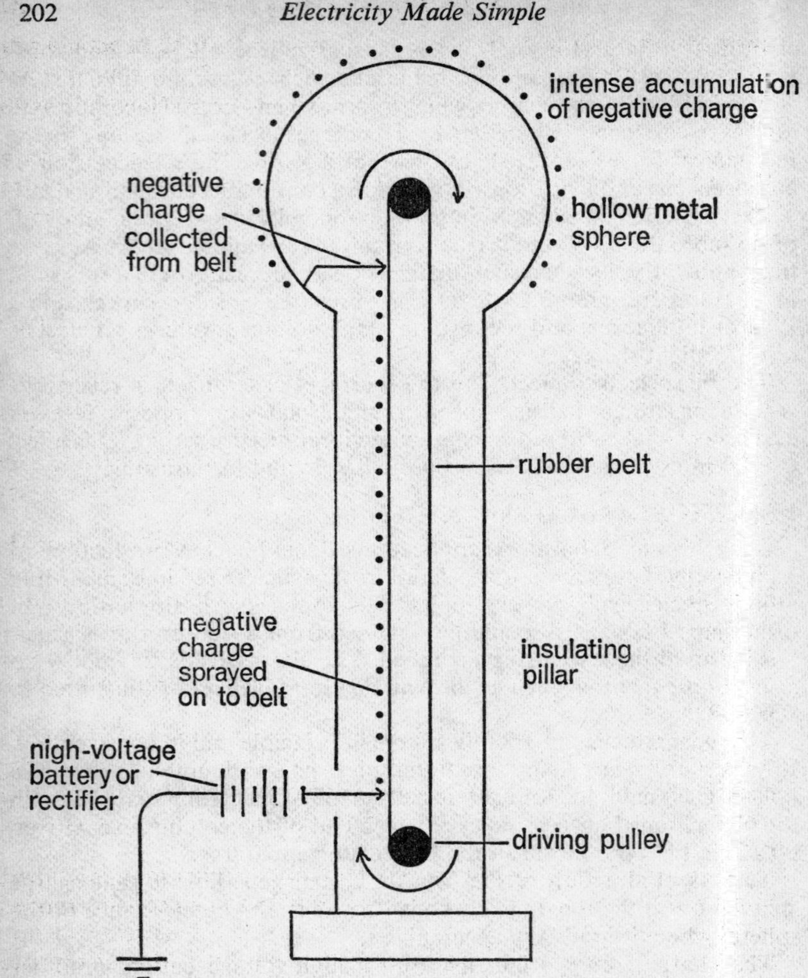

Fig. 132. Construction of the Van de Graaff electrostatic generator

repels electrons present in the comb, and these are provided with a conducting path to the inside of the sphere. Because charge will not remain on the inner surface of a hollow conductor, the electrons immediately pass to the *outside* of the sphere and have no tendency to return to the belt.

The build-up of charge on the sphere creates an increasing potential difference between the sphere and its surroundings. This p.d. is limited only by the quality of the insulation which prevents charge leaking away to earth. If the generator is surrounded by a tank containing dry air (or, preferably, sulphur hexafluoride gas) under pressure, a p.d. of the order of 4 MV can be achieved before the charge escapes violently to earth in the form of a spark.

When the generator is used for producing high-energy X-rays an X-ray tube is enclosed in the central column, parallel to the moving belt, with its cathode connected to the sphere and its anode connected to the earthed base.

WHAT NEXT?

Predicting future trends in a subject like electricity is a dangerous undertaking. One might, for example, prophesy the replacement of the Grid system by laser beams transmitting energy to entirely novel kinds of photoelectric generator in strategically placed substations: but today's Science Fiction is rarely tomorrow's hard fact. Yet there may well be some discovery waiting to revolutionize electrical engineering in the way that the advent of the transistor has changed the course of electronics.

On the basis of our present knowledge, the most likely advances would appear to lie in new methods of power generation, some of which have been outlined earlier in this chapter. In this connexion it is tempting to single out the *fuel cell* (see p. 87) as a promising field for development. A fuel cell would convert the chemical energy locked up in coal, oil, or natural gas into electricity without the intermediate stages of heat and mechanical energy.

A lightweight fuel cell would act as an ideal source of energy for driving motor vehicles, and the dream of a noiseless, fumeless, gearless electric car could at last be realized. Isolated dwellings might have their own fuel-cell generators, bringing them the benefits already available to consumers with a mains supply.

The output from a fuel cell is a low-voltage direct current, and therefore cannot be stepped up by means of a transformer. But modern semiconductor technology has made it possible for us to chop up a direct current into pulses (using a simple power-transistor or thyristor circuit) which can be transformed just like a true alternating current. Moreover, it is quite feasible to trigger the thyristors by means of a radio signal broadcast continuously from a central transmitter, so that the fuel-cell output could be converted into alternating current of precisely 50 Hz.

Thus the fuel cell promises to become an important new source of electricity, bringing widespread benefits to the motor-car-oriented urban society and the remote rural community alike.

TABLE OF NATURAL TRIGONOMETRIC FUNCTIONS

Degrees	Sin	Cos	Tan	Cot	
0° 00′	0·0000	1·0000	0·0000	—	90° 00′
10	029	000	029	343·8	50
20	058	000	058	171·9	40
30	0·0087	1·0000	0·0087	114·6	30
40	116	9999	116	85·94	20
50	145	999	145	68·75	10
1° 00′	0·0175	0·9998	0·0175	57·29	89° 00′
10	204	998	204	49·10	50
20	233	997	233	42·96	40
30	0·0262	0·9997	0·0262	38·19	30
40	291	996	291	34·37	20
50	320	995	320	31·24	10
2° 00′	0·0349	0·9994	0·0349	28·64	88° 00′
10	378	993	378	26·43	50
20	407	992	407	24·54	40
30	0·0436	0·9990	0·0437	22·90	30
40	465	989	466	21·47	20
50	494	988	495	20·21	10
3° 00′	0·0523	0·9986	0·0524	19·08	87° 00′
10	552	985	553	18·07	50
20	581	983	582	17·17	40
30	0·0610	0·9981	0·0612	16·35	30
40	640	980	641	15·60	20
50	669	978	670	14·92	10
4° 00′	0·0698	0·9976	0·0699	14·30	86° 00′
10	727	974	729	13·73	50
20	756	971	758	13·20	40
30	0·0785	0·9969	0·0787	12·71	30
40	814	967	816	12·25	20
50	843	964	846	11·83	10
5° 00′	0·0872	0·9962	0·0875	11·43	85° 00′
10	901	959	904	11·06	50
20	929	957	934	10·71	40
30	0·0958	0·9954	0·0963	10·39	30
40	987	951	992	10·08	20
50	0·1016	948	0·1022	9·788	10
6° 00′	0·1045	0·9945	0·1051	9·514	84° 00′
10	074	942	080	9·255	50
20	103	939	110	9·010	40
30	0·1132	0·9936	0·1139	8·777	30
40	161	932	169	8·556	20
50	190	929	198	8·345	10
7° 00′	0·1219	0·9925	0·1228	8·144	83° 00′
10	248	922	257	7·953	50
20	276	918	287	7·770	40
30	0·1305	0·9914	0·1317	7·596	30
40	334	911	346	7·429	20
50	363	907	376	7·269	10
8° 00′	0·1392	0·9903	0·1405	7·115	82° 00′
10	421	899	435	6·968	50
20	449	894	465	6·827	40
30	0·1478	0·9890	0·1495	6·691	30
40	507	886	524	6·561	20
50	536	881	554	6·435	10
9° 00′	0·1564	0·9877	0·1584	6·314	81° 00′
	Cos	Sin	Cot	Tan	Degrees

Degrees	Sin	Cos	Tan	Cot	
9° 00′	0·1564	0·9877	0·1584	6·314	81° 00′
10	593	872	614	197	50
20	622	868	644	084	40
30	0·1650	0·9863	0·1673	5·976	30
40	679	858	703	871	20
50	708	853	733	769	10
10° 00′	0·1736	0·9848	0·1763	5·671	80° 00′
10	765	843	793	576	50
20	794	838	823	485	40
30	0·1822	0·9833	0·1853	5·396	30
40	851	827	883	309	20
50	880	822	914	226	10
11° 00′	0·1908	0·9816	0·1944	5·145	79° 00′
10	937	811	974	066	50
20	965	805	0·2004	4·989	40
30	0·1994	0·9799	0·2035	4·915	30
40	0·2022	793	065	843	20
50	051	787	095	773	10
12° 00′	0·2079	0·9781	0·2126	4·705	78° 00′
10	108	775	156	638	50
20	136	769	186	574	40
30	0·2164	0·9763	0·2217	4·511	30
40	193	757	247	449	20
50	221	750	278	390	10
13° 00′	0·2250	0·9744	0·2309	4·331	77° 00′
10	278	737	339	275	50
20	306	730	370	219	40
30	0·2334	0·9724	0·2401	4·165	30
40	363	717	432	113	20
50	391	710	462	061	10
14° 00′	0·2419	0·9703	0·2493	4·011	76° 00′
10	447	696	524	3·962	50
20	476	689	555	914	40
30	0·2504	0·9681	0·2586	3·867	30
40	532	674	617	821	20
50	560	667	648	776	10
15° 00′	0·2588	0·9659	0·2679	3·732	75° 00′
10	616	652	711	689	50
20	644	644	742	647	40
30	0·2672	0·9636	0·2773	3·606	30
40	700	628	805	566	20
50	728	621	836	526	10
16° 00′	0·2756	0·9613	0·2867	3·487	74° 00′
10	784	605	899	450	50
20	812	596	931	412	40
30	0·2840	0·9588	0·2962	3·376	30
40	868	580	994	340	20
50	896	572	0·3026	305	10
17° 00′	0·2924	0·9563	0·3057	3·271	73° 00′
10	952	555	089	237	50
20	979	546	121	204	40
30	0·3007	0·9537	0·3153	3·172	30
40	035	528	185	140	20
50	062	520	217	108	10
18° 00′	0·3090	0·9511	0·3249	3·078	72° 00′
	Cos	Sin	Cot	Tan	Degrees

Degrees	Sin	Cos	Tan	Cot	
18° 00′	0·3090	0·9511	0·3249	3·078	72° 00′
10	118	502	281	047	50
20	145	492	314	018	40
30	0·3173	0·9483	0·3346	2·989	30
40	201	474	378	960	20
50	228	465	411	932	10
19° 00′	0·3256	0·9455	0·3443	2·904	71° 00′
10	283	446	476	877	50
20	311	436	508	850	40
30	0·3338	0·9426	0·3541	2·824	30
40	365	417	574	798	20
50	393	407	607	773	10
20° 00′	0·3420	0·9397	0·3640	2·747	70° 00′
10	448	387	673	723	50
20	475	377	706	699	40
30	0·3502	0·9367	0·3739	2·675	30
40	529	356	772	651	20
50	557	346	805	628	10
21° 00′	0·3584	0·9336	0·3839	2·605	69° 00′
10	611	325	872	583	50
20	638	315	906	560	40
30	0·3665	0·9304	0·3939	2·539	30
40	692	293	973	517	20
50	719	283	0·4006	496	10
22° 00′	0·3746	0·9272	0·4040	2·475	68° 00′
10	773	261	074	455	50
20	800	250	108	434	40
30	0·3827	0·9239	0·4142	2·414	30
40	854	228	176	394	20
50	881	216	210	375	10
23° 00′	0·3907	0·9205	0·4245	2·356	67° 00′
10	934	194	279	337	50
20	961	182	314	318	40
30	0·3987	0·9171	0·4348	2·300	30
40	0·4014	159	383	282	20
50	041	147	417	264	10
24° 00′	0·4067	0·9135	0·4452	2·246	66° 00′
10	094	124	487	229	50
20	120	112	522	211	40
30	0·4147	0·9100	0·4557	2·194	30
40	173	088	592	177	20
50	200	075	628	161	10
25° 00′	0·4226	0·9063	0·4663	2·145	65° 00′
10	253	051	699	128	50
20	279	038	734	112	40
30	0·4305	0·9026	0·4770	2·097	30
40	331	013	806	081	20
50	358	001	841	066	10
26° 00′	0·4384	0·8988	0·4877	2·050	64° 00′
10	410	975	913	035	50
20	436	962	950	020	40
30	0·4462	0·8949	0·4986	2·006	30
40	488	936	0·5022	1·991	20
50	514	923	059	977	10
27° 00′	0·4540	0·8910	0·5095	1·963	63° 00′
	Cos	Sin	Cot	Tan	Degrees

Degrees	Sin	Cos	Tan	Cot	
27° 00′	0·4540	0·8910	0·5095	1·963	63° 00′
10	566	897	132	949	50
20	592	884	169	935	40
30	0·4617	0·8870	0·5206	1·921	30
40	643	857	243	907	20
50	669	843	280	894	10
28° 00′	0·4695	0·8829	0·5317	1·881	62° 00′
10	720	816	354	868	50
20	746	802	392	855	40
30	0·4772	0·8788	0·5430	1·842	30
40	797	774	467	829	20
50	823	760	505	816	10
29° 00′	0·4848	0·8746	0·5543	1·804	61° 00′
10	874	732	581	792	50
20	899	718	619	780	40
30	0·4924	0·8704	0·5658	1·767	30
40	950	689	696	756	20
50	975	675	735	744	10
30° 00′	0·5000	0·8660	0·5774	1·732	60° 00′
10	025	646	812	720	50
20	050	631	851	709	40
30	0·5075	0·8616	0·5890	1·698	30
40	100	601	930	686	20
50	125	587	969	675	10
31° 00′	0·5150	0·8572	0·6009	1·664	59° 00′
10	175	557	048	653	50
20	200	542	088	643	40
30	0·5225	0·8526	0·6128	1·632	30
40	250	511	168	621	20
50	275	496	208	611	10
32° 00′	0·5299	0·8480	0·6249	1·600	58° 00′
10	324	465	289	590	50
20	348	450	330	580	40
30	0·5373	0·8434	0·6371	1·570	30
40	398	418	412	560	20
50	422	403	453	550	10
33° 00′	0·5446	0·8387	0·6494	1·540	57° 00′
10	471	371	536	530	50
20	495	355	577	520	40
30	0·5519	0·8339	0·6619	1·511	30
40	544	323	661	501	20
50	568	307	703	1·492	10
34° 00′	0·5592	0·8290	0·6745	1·483	56° 00′
10	616	274	787	473	50
20	640	258	830	464	40
30	0·5664	0·8241	0·6873	1·455	30
40	688	225	916	446	20
50	712	208	959	437	10
35° 00′	0·5736	0·8192	0·7002	1·428	55° 00′
10	760	175	046	419	50
20	783	158	089	411	40
30	0·5807	0·8141	0·7133	1·402	30
40	831	124	177	393	20
50	854	107	221	385	10
36° 00′	0·5878	0·8090	0·7265	1·376	54° 00′
	Cos	Sin	Cot	Tan	Degrees

Degrees	Sin	Cos	Tan	Cot	
36° 00′	0·5878	0·8090	0·7265	1·376	54° 00′
10	901	073	310	368	50
20	925	056	355	360	40
30	0·5948	0·8039	0·7400	1·351	30
40	972	021	445	343	20
50	995	004	490	335	10
37° 00′	0·6018	0·7986	0·7536	1·327	53° 00′
10	041	969	581	319	50
20	065	951	627	311	40
30	0·6088	0·7934	0·7673	1·303	30
40	111	916	720	295	20
50	134	898	766	288	10
38° 00′	0·6157	0·7880	0·7813	1·280	52° 00′
10	180	862	860	272	50
20	202	844	907	265	40
30	0·6225	0·7826	0·7954	1·257	30
40	248	808	0·8002	250	20
50	271	790	050	242	10
39° 00′	0·6293	0·7771	0·8098	1·235	51° 00′
10	316	753	146	228	50
20	338	735	195	220	40
30	0·6361	0·7716	0·8243	1·213	30
40	383	698	292	206	20
50	406	679	342	199	10
40° 00′	0·6428	0·7660	0·8391	1·192	50° 00′
10	450	642	441	185	50
20	472	623	491	178	40
30	0·6494	0·7604	0·8541	1·171	30
40	517	585	591	164	20
50	539	566	642	157	10
41° 00′	0·6561	0·7547	0·8693	1·501	49° 00′
10	583	528	744	144	50
20	604	509	796	137	40
30	0·6626	0·7490	0·8847	1·130	30
40	648	470	899	124	20
50	670	451	952	117	10
42° 00′	0·6691	0·7431	0·9004	1·111	48° 00′
10	713	412	057	104	50
20	734	392	110	098	40
30	0·6756	0·7373	0·9163	1·091	30
40	777	353	217	085	20
50	799	333	271	079	10
43° 00′	0·6820	0·7314	0·9325	1·072	47° 00′
10	841	294	380	066	50
20	862	274	435	060	40
30	0·6884	0·7254	0·9490	1·054	30
40	905	234	545	048	20
50	926	214	601	042	10
44° 00′	0·6947	0·7193	0·9657	1·036	46° 00′
10	967	173	713	030	50
20	988	153	770	024	40
30	0·7009	0·7133	0·9827	1·018	30
40	030	112	884	012	20
50	050	092	942	006	10
45° 00′	0·7071	0·7071	1·0000	1·000	45° 00′
	Cos	Sin	Cot	Tan	Degrees

APPENDIX

GLOSSARY OF TERMS USED IN ELECTRICITY

Italicized terms within individual definitions are defined elsewhere (in alphabetical order) in the Glossary.

A

A.C.—Abbreviation for *alternating current.*

ACCUMULATOR—Common name for *storage battery*, i.e. a battery which can be recharged by passing a current through it.

ALTERNATING CURRENT—Current which flows first in one direction then in the opposite direction repeatedly.

AMMETER—Instrument for measuring current in *amperes.*

AMPERE—Unit of *current*, defined as that current which, flowing in two infinitely long parallel straight conductors placed 1 metre apart in a vacuum produces a force of 2×10^{-7} newtons per metre length of the conductors.

AMPERE-TURN—Unit of magnetomotive force, equal to the m.m.f. produced in a coil of one turn when a current of 1 amp flows through it. An m.m.f. of 1 ampere-turn is equivalent to 1·2566 gilberts.

ANODE—The conductor to which *electrons* and negative *ions* are attracted in *electrolysis*, in gas discharge tubes, and in thermionic valves; it is connected to the positive side of the source of e.m.f.

APPARENT POWER—The product of effective current and effective e.m.f. in an a.c. circuit. To obtain the true or *real* power dissipated, the apparent power must be multiplied by the *power factor*. See *voltampere.*

ARMATURE—In general, any piece of (ferromagnetic) metal placed between the poles of a magnet; the term is applied particularly to the rotating core of a motor or generator.

B

BACK E.M.F.—The electromotive force induced by a changing current in a circuit that exhibits *self-inductance.*

BATTERY—Group of *cells* connected together in any combination.

BRUSH—Block of carbon (usually graphite) on which a rotating *commutator* rubs in order to lead current into or out of the rotating part of a motor or generator.

C

CAPACITANCE—Ability of a *capacitor* to store electric charge; it is defined by the ratio of the charge stored to the p.d. across the capacitor. Measured in farads, microfarads, and picofarads.

CAPACITIVE REACTANCE—That part of the *impedance* offered by a capacitor which does not dissipate power; for a purely capacitive circuit it is the ratio of the alternating p.d. to the current, and is inversely proportional to the frequency of the alternating p.d. Symbol X_c, measured in *ohms*.

CAPACITOR—Any system of conductors designed to store electrical charge; in practice, it consists essentially of two plates separated by an insulator.

CATHODE—The conductor which repels *electrons* and negative *ions* in *electrolysis*, in gas discharge tubes, and in thermionic valves; it is the most negative of any system of electrodes, and is connected to the negative side of the source of e.m.f.

CELL—Device in which chemical energy is converted into electrical energy.

CHARACTERISTIC—Strictly speaking, CHARACTERISTIC CURVE—graph showing the relationship between any two interdependent quantities, particularly the variation of the current through a conductor with the p.d. across it.

CIRCUIT—Any system of conductors which is capable of providing a closed path for electrons to flow along.

COEFFICIENT OF COUPLING—A measure of the relative efficiency with which magnetic *flux* produced in one coil links a second coil; if there is no leakage of flux between the coils the coefficient of coupling is unity. Symbol k, defined by the equation $M = k\sqrt{L_1L_2}$, where M is the mutual inductance between coils whose self-inductances are L_1 and L_2.

COERCIVE FORCE—Magnitude of the magnetic field needed to remove completely the magnetism remaining in a specimen that has been taken through a cycle of magnetization (see *hysteresis*).

COMMUTATOR—Rotating switch forming part of a motor or generator. It consists of a cylinder cut into segments which are connected to the ends of the armature winding; current enters and leaves the segments via fixed carbon *brushes* rubbing against them.

COMPOUND MOTOR—Electric motor in which the armature and field windings are connected partly in series and partly in parallel.

CONDUCTANCE—Reciprocal of *resistance*, a measure of the ease with which current passes through a conductor; symbol G, measured in reciprocal ohms or siemens (formerly the mho).

CONDUCTOR—Any device which offers a low resistance to the passage of current.

CORE LOSS—Energy wasted in an electrical machine (e.g. motor, transformer, or generator) owing to *hysteresis* and *eddy-current* heating.

COULOMB—Practical unit of electrical charge: a current of 1 ampere passing through a conductor for 1 second conveys 1 coulomb of charge.

COULOMB'S LAW—The force between two electrical charges is inversely proportional to the square of the distance between them, thus the force is a constant $\times \frac{q_1q_2}{r^2}$; the name is also applied to the force between two magnetic poles, in which case it is a constant $\times \frac{m_1m_2}{r^2}$.

CURRENT—Rate of flow of electric charges, particularly the rate of flow of electrons in a metallic conductor. Symbol I, measured in *amperes*.

CYCLE—Current variation (rise, fall, reverse, fall, rise) which repeats over and over again to form an alternating current.

D

D.C.—Abbreviation for *direct current.*

DEPOLARIZING AGENT—Chemical included in a *cell* to combine with hydrogen, removing it in the form of water, thereby preventing *polarization*; e.g. manganese dioxide in the dry cell.

DIELECTRIC—Electrical insulator, especially when forming part of a *capacitor.*

DIELECTRIC CONSTANT—Constant of proportionality in the relationship between the *capacitance* and dimensions of a capacitor; also called relative *permittivity.*

DIODE—Literally, any device having only two electrodes (anode and cathode); the term applies in particular to two-electrode thermionic valves and semiconductor devices—both of which are used as *rectifiers.*

DIRECT CURRENT—Flow of electrical charge whose direction does not vary with time.

E

EDDY CURRENT—Current induced in any large conductor which moves in a magnetic field or which is cut by a changing magnetic field.

EFFICIENCY—Ratio of the energy obtained from a device to the energy supplied to it; in the case of a transformer it is the ratio of the power in the secondary to the power in the primary.

ELECTROCHEMICAL EQUIVALENT—Mass of any substance liberated in *electrolysis* by one *coulomb* of electric charge.

ELECTRODE—Conductor through which current passes from a solid into a liquid, into a gas, or into a vacuum, and vice versa.

ELECTROLYSIS—Chemical decomposition brought about by passing a current through an *electrolyte.*

ELECTROLYTE—Solution of an acid, base, or salt which contains *ions* and therefore acts as an electrical conductor.

ELECTROMOTIVE FORCE—Voltage between the terminals of a generator or cell when it is on open circuit, i.e. delivering no current.

ELECTRON—Negatively charged particle present in all atoms; it is the carrier of electricity in metallic conductors. (There is also the positron, sometimes called the positive electron, which is equal in mass but opposite in the sign of its charge.)

E.M.F.—Abbreviation for *electromotive force.*

EQUIVALENT WEIGHT—Atomic weight of an element divided by its valency. Defined as the weight of the element that will combine with or displace one gram of hydrogen. Also known as CHEMICAL EQUIVALENT.

EXCITER—Self-contained d.c. generator which supplies the current needed to 'excite' the field windings of a large alternating-current generator.

F

FARAD—Unit of *capacitance* defined by the ratio of the charge (coulombs) stored on the plates of a capacitor to the p.d. (volts) between them.

FARADAY'S LAWS—(1) Laws of electrolysis: (*a*) the mass of any substance liberated in electrolysis is proportional to the quantity of charge passing; (*b*) the mass of substance liberated by a given current in a given time is proportional to the equivalent weight of that substance. (2) Law of electromagnetic induction: the e.m.f. induced in a circuit is proportional to the rate at which the magnetic flux cutting the circuit changes.

FERROMAGNETIC MATERIAL—Iron, cobalt, nickel, and their alloys which are strongly attracted by a magnet; i.e. materials whose magnetic *permeability* is very high relative to that of empty space.

FIELD—Region in which the effects of a magnet, electric charge, etc., can be detected. The strength or intensity of a field is defined by the force exerted on a known magnetic pole or known electric charge situated at the point in question.

FIELD WINDING—Conductors which create the necessary magnetic field in an electric motor or generator.

FLUX DENSITY—Magnetic flux passing at right angles through a given area (usually one square metre or one square centimetre).

FREE ELECTRON—Electron that is bound to no particular atom, and can therefore move from atom to atom in a conductor, thus forming a current of electricity.

FREQUENCY—the number of times an oscillation, particularly an alternating current, repeats itself in one second; hence it is measured as the number of cycles per second (c/s), an alternative unit being the hertz (Hz).

FULL-WAVE RECTIFIER—Device that converts both the positive and the negative a.c. half-cycles into a one-way (direct) current.

G

GALVANOMETER—Device for detecting and (usually) measuring small *currents.*

H

HALF-WAVE RECTIFIER—Device that offers low resistance to either the positive or the negative (but not both) half-cycles in an alternating current.

HENRY—Practical unit of *inductance*; defined by the inductance of a circuit in which an e.m.f. of one volt is induced by a current changing at the rate of one ampere per second. Symbol H.

HERTZ—Unit of *frequency* in the SI unit system; equal to the cycle per second, symbol Hz.

HYSTERESIS—Lagging of the *flux* induced in a magnetic material behind the magnetizing *field.* Thus when a magnetic field is varied in a repeating pattern (e.g. increase, reduce, reverse, increase, etc.) the flux induced in a specimen in the field varies in the same pattern, but its maxima and minima occur after those of the field.

HYSTERESIS LOSS—Energy converted to heat in a magnetic material subjected to a magnetic field that varies in a repeating pattern or cycle.

I

IMPEDANCE—Opposition presented by a circuit to the passage of *alternating current*; measured by the ratio of alternating voltage/alternating current. Symbol Z, measured in ohms.

INDUCED E.M.F.—Voltage created in a circuit by a changing magnetic flux. The current resulting from an induced e.m.f. is called INDUCED CURRENT.

INDUCTION MOTOR—Electric motor in which the magnetic field of the rotor is set up by induced currents; there need be no electrical connexion to the rotor.

INDUCTIVE REACTANCE—That part of the *impedance* offered by an inductor which does not dissipate power; for a purely inductive circuit it is the ratio of the alternating p.d. to the current, and is directly proportional to the frequency of the alternating p.d. Symbol X_L, measured in ohms.

INDUCTOR—A component that is included in a circuit because it has INDUCTANCE, the property whereby voltage is created in it by a changing magnetic flux. In practice, an inductor is a coil of wire, which may have a solid core of ferromagnetic material or, in the case of radio-frequency currents, an empty air-cored former.

INTERNAL RESISTANCE—Resistance of a battery, cell, or generator.

ION—Atom or molecule that has become electrically charged by either gaining or losing electrons.

IONIZATION—Any process by which an atom or molecule is converted into an ion.

I^2R LOSS—Energy converted into heat in the windings of a motor, generator or transformer; measured in units of power, i.e. energy loss per second.

J

JOULE—Practical unit of energy, defined by the work done when a force of 1 newton acts through a distance of 1 metre in the direction of the force.

JOULE'S LAW—If a current I amps flows through a resistance R ohms for a period of t seconds the energy converted into heat is I^2Rt joules.

K

KILO—Prefix indicating thousand: e.g. kilowatt (kW) = 1,000 W; kilovolt (kV) = 1,000 V.

KILOWATT-HOUR—Practical unit of energy defined by a power of 1,000 W operating for a period of 1 hour; equivalent to 3,600,000 joules. Symbol kWh.

KIRCHHOFF'S LAWS—(1) Total current flowing towards a junction equals the total current leaving that junction. (2) The sum of the $I \times R$ products in any closed circuit equals the total e.m.f. in that circuit.

L

LAMINATIONS—Thin sheets into which the cores of electrical machines are divided in order to reduce *eddy currents.*

LENZ'S LAW—The direction of an induced current is always such as to oppose the motion or change causing it.

LINES OF FORCE—Imaginary lines showing the direction of a *field.* A line of magnetic force is the path an isolated N pole would take if free to travel.

M

MAGNETIC DOMAIN—Group of some 10^{15} atoms in a *ferromagnetic* material where the electrons spin in the same direction, producing an intense local magnetic field.

MAGNETIC FIELD—Region in which magnetic effects are present; the strength or intensity of the field, also called MAGNETIZING FORCE, is the force exerted on a magnetic pole placed therein.

MAGNETIC FLUX—A measure of the magnetism 'flowing' across a given area; its unit is the weber (Wb) defined as the flux which, when cut at a uniform rate by a conductor in 1 second, generates an e.m.f. of 1 volt. The C.G.S. unit of flux is the maxwell = 10^{-8} webers.

MAGNETIC FLUX DENSITY—Amount of magnetic flux per square metre. Symbol B. Its unit is the weber per square metre, defined as the density of a magnetic field where a force of 1 newton is exerted on a current of 1 ampere flowing at right angles to the field.

MAGNETOMOTIVE FORCE—Work done in taking a magnetic pole around a closed magnetic circuit. It is analogous to electromotive force. Symbol F, measured in *ampere-turns* (or gilberts in the C.G.S. system).

MAXWELL—C.G.S. unit of *magnetic flux*.

MEGA—Prefix indicating million: e.g. megawatt (MW) = 1,000,000 W; megavolt (MV) = 1,000,000 V.

MICRO—Prefix indicating millionth: e.g. microamp (μA) = 10^{-6} A; microfarad (μF) = 10^{-6} farads.

MILLI—Prefix indicating thousandth: e.g. millivolt (mV) = 10^{-3} V.

M.M.F.—Abbreviation for *magnetomotive force*.

MOTOR—Any device in which electrical energy is converted into energy of motion.

MUTUAL INDUCTANCE—Property of two electrically separate circuits by which a changing current in one induces an e.m.f. in the other. Measured in *henrys*.

N

n-TYPE SEMICONDUCTOR—Semiconducting material, e.g. germanium or silicon, that has been 'doped' with an impurity, e.g. arsenic, to give it an excess of electrons. The *n* stands for negative.

O

OHM—Unit of resistance, defined as the resistance of a conductor in which a current of 1 ampere generates heat at the rate of 1 joule per second. Symbol Ω.

OHM'S LAW—The current in a conductor is directly proportional to the potential difference across the ends of the conductor, provided that the physical conditions such as temperature remain constant.

P

PARALLEL CONNEXION—Method of joining two or more components so that the same p.d. exists across each of them; cells and other sources of e.m.f. are in parallel when there is one conductor connecting all the positive terminals and another joining all the negative terminals.

P.D.—Abbreviation for *potential difference*.

PELTIER EFFECT—Heating of one junction and cooling of the other when a current flows in a circuit of two dissimilar metals.

PERMEABILITY—Ratio of the magnetic flux density B in a material to the intensity H of the magnetic field in which it is situated. Symbol μ (μ =

B/H). The permeability of free space is $\mu_0 = 4\pi \times 10^{-7}$ (in rationalized M.K.S. units). Relative permeability μ_r is the ratio μ/μ_0.

PERMITTIVITY, RELATIVE—Also known as dielectric constant—ratio of the electrical flux density produced in a dielectric by a given electric field to the flux density produced in free space by the same field.

PHASE ANGLE—Angle between two *vectors* representing oscillating quantities, particularly vectors representing alternating current and voltage. Symbol θ.

PICO—Prefix indicating billionth (10^{-12}): e.g. picofarad (pF) = 1 micro-microfarad = 10^{-12} F.

POLARIZATION—Defect of a cell in which the positive plate becomes coated with hydrogen.

POLE—(1) Either terminal of a battery or cell. (2) One end of a magnet, particularly the electromagnet system employed in motors and generators.

POTENTIAL—Electrical potential is a measure of the work needed to bring one unit of charge from an infinitely great distance to the point in question.

POTENTIAL DIFFERENCE—Work needed to tranfer one unit of charge between the two points whose difference of potential is being measured. Symbol V. Measured in volts.

POTENTIAL ENERGY—Energy which any object possesses by virtue of its position.

POTENTIOMETER—(1) Type of variable resistor. (2) Circuit for comparing one p.d. with another.

POWER—Rate at which energy is converted from one form to another. A power of 1 joule per second is 1 watt.

POWER FACTOR—Ratio of the *real power* to the *apparent power* in an a.c. circuit. It is the cosine of the *phase angle*, written cos θ.

p-TYPE SEMICONDUCTOR—Semiconducting material, e.g. germanium or silicon, that has been 'doped' with an impurity, e.g. indium, to give it an excess of 'holes', i.e. a deficit of electrons. The *p* stands for positive.

Q

Q FACTOR—An indication of the 'quality' of a resonant circuit, obtained by comparing the energy stored with the rate of energy loss. It is measured by the ratio of the reactance of either the inductor or the capacitor at the resonant frequency to the total resistance of the circuit; thus $Q = Z_L/R = Z_C/R$.

R

RADIAN MEASURE—System of angular measurement in which the size of an angle is defined by the ratio of the length of the circular arc subtending the angle to the radius of the arc. Thus a complete circle of radius r is an arc of length $2\pi r$, and subtends an angle of $2\pi r/r = 2\pi$ radians. To convert from radians to degrees, multiply by 57·3.

REACTIVE POWER—Maximum rate at which stored energy surges into and out of a circuit, coming from the source and then all going back to it; reactive power is alternately 'stored' and returned by the *capacitors* and *inductors* of an a.c. circuit.

REACTANCE—That part of a circuit's *impedance* which is non-resistive, i.e.

in which energy is not converted to heat. Symbol X. See also *capacitive reactance* and *inductive reactance.*

REAL POWER—Power dissipated in an a.c. circuit, i.e. the rate at which energy is converted in the resistive components. Real power, also called true power, is the product of the *apparent power* and the *power factor.* Measured in watts.

RECTIFICATION—The conversion of alternating current into direct current. Any device which will perform this function is called a RECTIFIER.

RELUCTANCE—Ratio of the magnetomotive force to the flux in a magnetic circuit. Symbol R. Reluctance is the magnetic analogue of electrical resistance.

REMANENCE—Flux remaining in a specimen after the magnetizing field has been reduced to zero. Also called RESIDUAL MAGNETISM.

RESISTANCE—The property of a conductor which opposes the flow of current through it and converts electrical energy into heat. Defined by the ratio of the p.d. across the ends of a conductor to the current flowing. Symbol R, measured in ohms.

RESISTIVITY—Resistance of a conductor whose length is one unit and whose area of cross-section is one unit. Also called *specific resistance.* Symbol ρ, measured in ohm-metres (M.K.S. system) or ohm-cm (C.G.S. system).

RESISTOR—Any component included in a circuit because it possesses the property of *resistance.*

RESONANCE—Condition which occurs in a circuit containing inductance and capacitance when, at a particular frequency called the RESONANT FREQUENCY, the current is maximum and the impedance is minimum (for a series circuit) or the impedance is maximum and the current is minimum (for a parallel circuit). In either case the *inductive reactance* of the circuit is equal to the *capacitive reactance.*

RHEOSTAT—A type of variable resistor.

R.M.S.—Abbreviation for *root-mean-square.*

ROOT-MEAN-SQUARE—Magnitude of an alternating current that has the same heating effect as a numerically equal direct current; it is found by taking the square root of the average of the $(\text{current})^2$. Root-mean-square values are also called effective values, and apply to alternating voltages as well as current. Peak or maximum instantaneous value is equal to $\sqrt{2} \times$ r.m.s. value.

ROTOR—Revolving system of conductors in a motor or generator.

S

SATURATION—The condition of a *ferromagnetic* material when all its *magnetic domains* are aligned.

SEEBECK EFFECT—The appearance of a thermoelectric e.m.f. in a circuit of two dissimilar metals when their junctions are at different temperatures.

SELF-INDUCTANCE—Property of a circuit (typically containing a coil) in which an e.m.f. is induced by a change of current. Symbol L, measured in henrys.

SEMICONDUCTOR—Material such as germanium whose electrical conducting properties are intermediate between those of metals and those of insulators. Small quantities of a suitable impurity have a profound effect on the conducting properties of these materials, turning them into either p-type or n-type semiconductors.

SENSITIVITY—Of a meter, the ratio of the deflection produced to the magnitude of the current, voltage, etc., causing it.

SERIES CONNEXION—Method of joining two or more components so that the same current flows through each in turn.

SERIES MOTOR—Electric motor in which the armature and field windings are joined in series.

SHUNT—In general, any circuit component connected in parallel with another; in particular, a small resistor in parallel with a *galvanometer*.

SHUNT MOTOR—Electric motor in which the armature and field windings are joined in parallel.

SKIN EFFECT—Effect whereby alternating currents, particularly those of high frequency, tend to flow near the surface of a conductor.

SLIP RINGS—Pair of metal rings fixed to the shaft of a motor or generator enabling alternating current to enter and leave the rotor; the external circuit is connected to *brushes* rubbing against the rings.

SOLENOID—Coil whose length is great compared with its diameter; such a coil is generally used for producing a uniform magnetic field.

SPECIFIC RESISTANCE—Also called *resistivity*, the resistance of a conductor whose length is one unit and whose area of cross section is also one unit. Symbol ρ, measured in ohm-metres or ohm-cm.

SQUIRREL-CAGE MOTOR—Very popular type of *induction motor* in which the rotor is constructed from thick conducting rods arranged like the bars of a circular cage.

STATOR—Stationary system of conductors in a motor or generator.

STORAGE BATTERY—Battery of cells in which the chemical energy can be restored by passing a current through them in the appropriate direction.

SUPERCONDUCTIVITY—Property of certain metals and alloys whose electrical *resistance* is almost nil at temperatures near the absolute zero.

T

TANK CIRCUIT—Resonant circuit comprising a *capacitor* in parallel with an *inductor*. Used in the tuning stage of a radio receiver since it is 'selective', i.e. offers a greater impedance to currents alternating at the *resonant frequency* than to any other currents.

TEMPERATURE COEFFICIENT OF RESISTANCE—Fractional change in the *resistance* of a material for each degree of temperature change. Symbol α. If the resistance changes from R_0 to R_t when the temperature increases from 0° to t° C, the temperature coefficient of resistance is $\alpha = (R_t - R_0)/R_0t$ per deg C.

TESLA—Unit of magnetic *flux density* in the SI unit system; equal to 1 weber per square metre.

THERMISTOR—Device whose resistance decreases as its temperature rises; i.e. one which exhibits a negative *temperature coefficient of resistance*.

THERMOCOUPLE—Junction between two dissimilar metals; when the junction is heated or cooled relative to the rest of the circuit, a THERMOELECTRIC E.M.F. is developed across it.

THERMOPILE—Several *thermocouples* joined in series and arranged so that alternate junctions lie together in order to make their individual e.m.f.s additive.

THREE-PHASE SUPPLY—Source of electrical power consisting of three separate alternating voltages out of phase with one another by one-third of a cycle, i.e. 120°.

THYRISTOR—Transistor equivalent of the gas-filled valve, this device is a *semiconductor* (silicon) *rectifier* incorporating a third terminal or gate whose voltage controls the conditions under which the rectifier will conduct. Formerly called a silicon-controlled rectifier (s.c.r.).

TORQUE—Turning effect of a force. Measured in newton-metres, dyne-cm, pound-ft, etc.

TRANSFORMER—Device for changing the voltage of an a.c. supply; in particular, an arrangement of two *inductors* wound on a common core so that the magnetic flux created in one cuts the second and sets up an *induced e.m.f.* in it.

U

UNIVERSAL MOTOR—Electric motor comprising a stator, wound rotor, and commutator, which can be operated from both a.c. and d.c. supplies.

V

VECTOR—Any quantity having direction as well as magnitude; the term is also applied to a line drawn to represent such a quantity.

VOLT—Practical unit of *electromotive force* and *potential difference*. The p.d. between two points on a conductor is defined as 1 volt when a current of 1 amp passing between them dissipates a power of 1 watt.

VOLTAMPERE—Unit in which *apparent power* is measured, thus distinguishing it from real power measured in watts. The product of an alternating current I amps (r.m.s.) and an alternating voltage V volts (r.m.s.) is VI voltamperes. Symbol VA.

VOLTMETER—Instrument for measuring potential differences.

W

WATT—Practical unit of *power*; defined as a rate of converting energy or of doing work of 1 *joule* per second. In electrical terms it is the rate at which energy is dissipated by a current of 1 amp flowing across a p.d. of 1 volt. Symbol W.

WEBER—Unit of *magnetic flux* defined as the flux which, when cut at a uniform rate by a conductor in 1 second, generates an e.m.f. of 1 volt. Symbol Wb.

WESTON CELL—Cell comprising mercury and cadmium amalgam electrodes, which when constructed to a standard specification has an e.m.f. of exactly 1·0186 volts at 20° C. It is used as reference against which voltages are checked.

WHEATSTONE BRIDGE—Apparatus for measuring *resistance*. It consists of a circuit of four arms: the unknown resistance, a standard resistor, and two resistors whose ratio is known. In the 'metre bridge' the two ratio resistors are replaced by a length of resistance wire of uniform cross-section.

REVISION QUESTIONS

Most of the following questions should present little difficulty, but one or two need to be approached carefully. Their main purpose is to help the reader discover which areas of the text have been properly understood and which—if any—call for further study. They are not intended to reflect the level or type of question used in any specific examination. Answers are provided on p. 224.

1. Supply suitable words to fill the blanks in the following passage. Electrons are minute .... charged particles which move in orbits around the central .... of an atom. Free .... exist independently of the atoms from which they have been dislodged. An electric .... is an organized drift of free electrons: the size of the current is the .... of flow of electrons. Metals contain relatively .... numbers of free electrons, and are therefore .... conductors of electricity. .... materials contain very few free electrons.

2. (*a*) If a conductor obeys Ohm's law, what kind of graph would result from plotting the current through the conductor against the potential difference between its ends?

(*b*) How could the *resistance* of the conductor be deduced from the graph?

(*c*) Name two conducting systems which do *not* obey Ohm's law.

3. When a resistance of R ohms is connected across the terminals of a battery it is found that a current of I amperes passes through it, and the potential difference between the terminals is V volts. If the battery has an e.m.f. of E volts and an internal resistance of r ohms, what is the relationship (*a*) between V and E (*b*) between r and R?

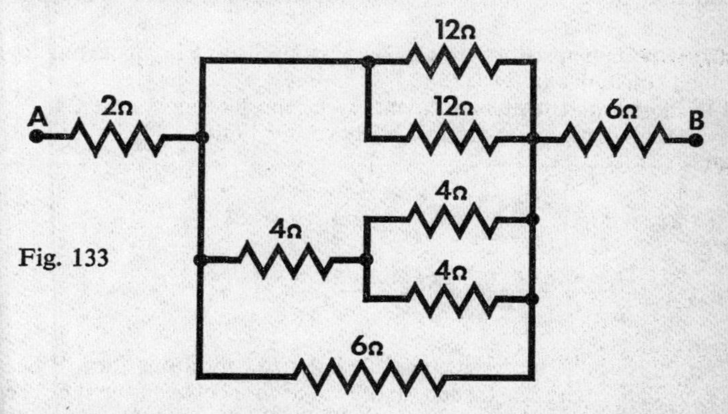

Fig. 133

4. A single resistor connected between A and B could exactly replace the network shown in Fig. 133. What is its value?

5. Three batteries are joined in series. Their e.m.f.s and internal resistances are as indicated in Fig. 134. What readings would you expect to see on a voltmeter connected across each battery in turn?
(Hint: complete Question 3 before tackling this problem.)

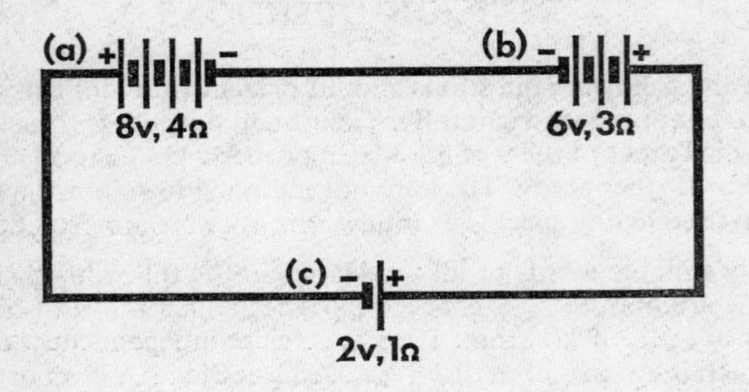

Fig. 134

6. A tungsten light bulb is labelled 240 V, 100 W.

(*a*) Calculate the resistance of the filament.
(*b*) When measured at room temperature the resistance is found to be 60 Ω. How do you account for the discrepancy between the calculated resistance and the measured resistance?
(*c*) Assuming that the filament operates at 2000°C above room temperature, what is the temperature coefficient of resistance of tungsten approximately?

7. Two wires, A and B, are identical in every respect, and their resistances are of course in the ratio 1 : 1. Wire B is now cut into 20 equal lengths which are twisted together in the form of a cable. What is the new ratio of the two resistances?

8. (*a*) How can resistors of 12 Ω, 18 Ω and 36 Ω be arranged to give a combined resistance of 24 Ω?
(*b*) If the arrangement is accidentally connected across the 240 V mains supply, which component is most likely to burn out?

9. (*a*) $Xl_2 = Sl_1$ (*b*) $\frac{X}{l_1} = \frac{S}{l_2}$ (*c*) $Xl_1 = Sl_2$

(*d*) $\frac{X}{S} = \frac{l_1}{l_2}$ (*e*) $X = \frac{Sl_1}{l_2}$

The formulas above represent possible balance conditions for a Wheatstone (metre) bridge, where l_1 is the length of bridge wire between the sliding contact and the (unknown) resistance whose value is X, and l_2 is the length of bridge wire between the sliding contact and the (standard) resistor whose value is S. Which formula is *not* valid?

10. (*a*) An electric kettle containing 2 kilograms of water takes 4 minutes to reach boiling point when connected to a 240 V supply. Assuming that the initial temperature is 40°C, how much energy is absorbed by the water? (Specific heat capacity of water is 4200 J/kg °C, i.e. 4200 joules raise the temperature of 1 kilogram of water by 1°C.)

(*b*) Neglecting all heat losses, how much electrical energy is converted into heat?

(*c*) What is the rate of energy conversion *per second*?

(*d*) What is the power rating of the kettle?

(*e*) What current is taken from the mains?

11. If the price of electricity is 2·5 p per kWh, what is the cost of charging a 12 V car battery at 2 A for 16 hours 40 minutes? Neglect all wasted energy.

12. Since fuses do nothing to improve the efficiency of an appliance, why are they so widely used? Explain briefly how a fuse operates.

13. Fill the blanks in the following passage.

The domains in an ordinary lump of iron are . . . . in all directions, and their . . . . effects cancel out. When the iron is placed near a powerful magnet (i.e. in a magnetic . . . .) the domains begin to rotate, a few at a time, to align themselves with the . . . . This is a gradual process that depends on the . . . . of the applied field and on the length of time for which it is present. Eventually, when all the domains have jumped into line, magnetization is complete and the iron is described as magnetically . . . .

14. Sketch the magnetic field of flux pattern produced by an electric current in

(*a*) a straight wire

(*b*) a long coil of wire.

Indicate in each case the direction of the (conventional) current and the direction of the magnetic flux.

15. By means of a labelled diagram illustrate the mechanism of *either* an electromagnetic relay *or* an electric bell.

16. A tray made of cupro-nickel is to be electroplated with silver.

(*a*) Is the tray connected to the positive or the negative terminal of a suitable voltage source?

(*b*) Suggest a possible material for the electrolyte.

(*c*) What carries the current in the electrolyte?

(*d*) What is the advantage of using a *small* current?

(*e*) What is the disadvantage of using a small current?

17. (*a*) Sketch what you would see if a dry cell is cut vertically through the middle. Label the various components.

(*b*) What is meant by 'polarization' of a cell? How is this defect overcome in the dry cell?

18. The following table lists certain features of a lead–acid storage battery. Fill in the blanks.

		Fully charged	*Fully discharged*
(*a*)	Colour of pos. plates	brown	—
(*b*)	Colour of neg. plates	—	white
(*c*)	Specific gravity of acid	1·25	—

19. (*a*) capacitance, (*b*) charge, (*c*) conductance, (*d*) current, (*e*) electromotive force, (*f*) energy, (*g*) force, (*h*) frequency, (*j*) inductance, (*k*) magnetic field strength, (*l*) magnetic flux, (*m*) magnetic flux density, (*n*) potential difference, (*o*) power, (*p*) work.

Select from the following list an appropriate unit for measuring each of the above quantities. Note that some units may be used twice and some not at all.

(i) ampere (ii) ampere-turn per metre
(iii) coulomb (iv) farad
(v) gilbert (vi) henry
(vii) hertz (viii) joule
(ix) newton (x) ohm
(xi) siemens (xii) tesla
(xiii) volt (xiv) watt
(xv) weber

20. The ampere is defined internationally as the current in a long straight wire in a vacuum which exerts a force of exactly x newtons on a parallel straight wire 1 metre away carrying the same current. Which of the following is the correct value of x ?

(*a*) 10^{-7} (*b*) 2×10^{-7} (*c*) 3×10^{-7} (*d*) 4×10^{-7} (*e*) 5×10^{-7}

21. (*a*) What is the potential difference between two points if a current of 5 amps passing between them dissipates a power of 1000 watts?

(*b*) What is the potential difference between two points if 200 joules of work is done in conveying 1 coulomb of electricity between them?

22. During a thunderstorm a stream of electrons rushes down a vertical lightning conductor. If the earth's magnetic field at this point is horizontal and its direction is south to north, what is the direction of the force on the conductor?

23. (*a*) Sketch the essential features of a moving-coil galvanometer.

(*b*) What is the chief factor governing the sensitivity of a moving-coil galvanometer?

24. A moving-coil galvanometer gives a full-scale deflection for a current of 2 mA. If the resistance of the instrument is 20 ohms, what external resistor will enable it to measure currents of up to 2 A ?

25. What external resistor will enable the galvanometer in Question 24 to measure voltages of up to 10 V?

26. (*a*) Briefly explain the purpose of a *commutator* in a d.c. motor.
(*b*) How can the *direction* of rotation of a d.c. motor be altered?

27. (*a*) Why does an electric motor use a greater current when starting than when running at full speed?
(*b*) Give one useful feature of a *shunt-wound* motor.
(*c*) Give one useful feature of a *series-wound* motor.

28. Which of the modifications listed below would *not* increase the e.m.f. obtained from a simple generator?
(*a*) Altering the speed of rotation.
(*b*) Using a more powerful field magnet.
(*c*) Replacing the slip rings by a commutator.
(*d*) Adding more turns to the coil.
(*e*) Winding the coil on a soft-iron core.

29. (*a*) Complete the following statement:
'An induced current is always in such a direction . . . .'
(*b*) Whose name is generally associated with this statement? (i) M. Faraday, (ii) J. P. Joule, (iii) H. E. Lenz, (iv) J. C. Maxwell, (v) G. S. Ohm

30. Why are the armatures of electrical machinery fabricated from thin steel stampings instead of being cast as a solid block?

31. A slide projector contains a 12 V, 150 W lamp and a transformer enabling it to be used with a 240 V a.c. supply.
(*a*) If the secondary (low-voltage) winding of the transformer has 100 turns, how many turns has the primary?
(*b*) If there is no loss of flux in the transformer, what current is drawn from the supply?
(*c*) If the transformer's efficiency is 90 per cent, what current is drawn from the supply?

32. (*a*) 120 kilowatts of power is transmitted at 240 volts via a cable whose resistance is 0·1 ohm. How much power is wasted as heat in the cable?
(*b*) 120 kilowatts of power is transmitted at 240 kilovolts via the same cable. How much power is wasted?
(*c*) Why is alternating current preferable to direct current for the distribution of power?

33. (*a*) What is the 'effective' or 'r.m.s.' value of an alternating current?
(*b*) What is the relationship between the r.m.s. value E of an alternating voltage and the peak value E_{max}?

34. The hot-wire ammeter has the advantage that it can be used without modification in both a.c. and d.c. circuits. Give *two* disadvantages of this instrument compared with the moving-coil type.

35. Sketch the circuit of a *full-wave* rectifier which does not require a centre-tapped transformer.

36. A capacitor consisting of two parallel plates is connected to a battery. Which of the following changes will *reduce* the quantity of charge stored by the capacitor?

(*a*) Increasing the area of the plates.
(*b*) Increasing the separation of the plates.
(*c*) Inserting a sheet of glass between the plates.
(*d*) Adding a second battery in series with the first.

37. A 32 μF capacitor is combined with a second 32 μF capacitor (*a*) in series, (*b*) in parallel. What is the total capacitance in each case?

38. The passage below refers to ideal components in an a.c. circuit. Fill in the blanks.

(i) For the capacitor the voltage curve reaches its peak values 90° . . . . the current curve; in other words, the current . . . . the applied voltage by 90° or . . . . radians.
(ii) For the inductor the current curve reaches its peak values 90° . . . . the voltage curve; in other words, the current . . . . the voltage by 90°.

39. (*a*) What unit is used for measuring the *real* power in an a.c. circuit?
(*b*) What unit is used for measuring the *apparent* power in an a.c. circuit?
(*c*) What name is given to the ratio $\frac{\text{real power}}{\text{apparent power}}$?
(*d*) What is the numerical value of this ratio when the *phase angle* is zero?

40. Explain why, in a Van de Graaff electrostatic generator, (*a*) the teeth of the lower comb become strongly charged, and (*b*) charge collected by the upper comb passes immediately to the outside of the sphere.

ANSWERS

1. See p. 10
2. (*a*) A straight line passing through the origin
   (*b*) The gradient or slope of the graph is I/V, hence 1/gradient is $V/I = R$
   (*c*) See p. 8
3. (*a*) $V = E - Ir$
   (*b*) $r = \frac{E - IR}{I}$ or $r = \frac{E}{I} - R$
4. 10 ohms
5. (*a*) 6 V (*b*) 7·5 V (*c*) 1·5 V
6. (*a*) 576 ohms
   (*b*) Resistance of tungsten increases with temperature; data used in the calculation relate to the filament when it is white hot
   (*c*) 0·0043 per °C
7. 400 : 1 or 1 : 0·0025

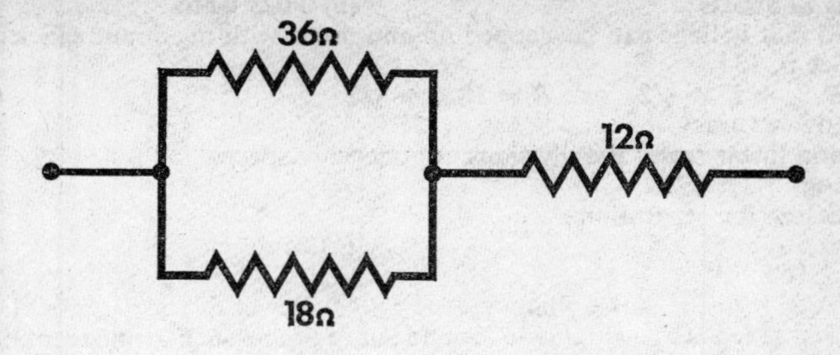

8. (*a*) As in the diagram above
   (*b*) The 12 Ω resistor, because it has the greatest power dissipation (1200 W)
9. $Xl_1 = Sl_2$
10. (*a*) 504 kJ (*b*) 504 kJ (*c*) 2100 J/s
    (*d*) 2100 W or 2·1 kW (*e*) 8·75 A
11. One penny
12. See p. 45
13. See p. 51
14. (*a*) See Fig. 30, p. 55 (*b*) See Fig. 32, p. 57
15. (*a*) See Fig. 36, p. 61 (*b*) See Fig. 37, p. 62
16. (*a*) Negative
    (*b*) Water containing any soluble silver salt, e.g. silver nitrate, but commercial users prefer the cyanide
    (*c*) Ions
    (*d*) It gives a firm, even deposit
    (*e*) The process takes longer
17. (*a*) See Fig. 48, p. 83
    (*b*) See p. 82; hydrogen ions are converted (oxidized) to water by the manganese dioxide
18. (*a*) White (*b*) Grey (*c*) Less than 1·2
19. (*a*) iv (*b*) iii (*c*) xi
    (*d*) i (*e*) xiii (*f*) viii
    (*g*) ix (*h*) vii (*j*) vi
    (*k*) ii (*l*) xv (*m*) xii
    (*n*) xiii (*o*) xiv (*p*) viii
20. 2×10^{-7}
21. (*a*) 200 volts (*b*) 200 volts
22. Horizontally, east to west
23. (*a*) See Fig. 53, p. 94
    (*b*) Intensity of the magnetic flux: in practice this depends on the efficiency of the permanent magnet
24. 0·02 Ω in parallel
25. 4980 Ω in series
26. (*a*) See p. 99
    (*b*) By reversing *either* the current in the armature *or* the current in the field windings (if used)
27. (*a*) See p. 99 (*b*) Constant speed (*c*) High starting torque
28. Replacing the slip rings by a commutator
29. (*a*) See p. 109 (*b*) Lenz
30. To reduce the energy wasted in setting up eddy currents
31. (*a*) 2000 (*b*) 0·625 A (*c*) 0·694 A

32. (*a*) 25 kilowatts (*b*) 0·025 watts
(*c*) So that voltage can be stepped up and down with maximum efficiency
33. (*a*) See p. 134
(*b*) $E_{max} = E \times \sqrt{2}$ or $E = E_{max} \div \sqrt{2}$
34. (*a*) Low accuracy
(*b*) Non-linear scale: the divisions are unevenly spaced
35. See Fig. 84, p. 141
36. Increasing the separation
37. (*a*) 16 μF (*b*) 64 μF
38. See p. 160
39. (*a*) Watt (*b*) Voltampere (*c*) Power factor (*d*) 1·0
40. (*a*) Charge tends to accumulate around sharp points such as the teeth of a metal comb
(*b*) Owing to mutual repulsion, charge will not remain on the inner surface of any hollow conductor

Index